BIRMINGHAM CITY
University

Envir

Please
remember to
return or
renew on time
to avoid fines

Renew/check due dates via
www.bcu.ac.uk/library

BORDERLINES

A BOOK SERIES CONCERNED WITH REVISIONING GLOBAL POLITICS
Edited by David Campbell and Michael J. Shapiro

For more books in the series, see p. vi.

Environmental Security

SIMON DALBY

BORDERLINES, VOLUME 20

 University of Minnesota Press

Minneapolis

London

Parts of chapters 1 and 8 were published in *Geopolitical Change and Contemporary Security Studies: Contextualizing the Human Security Agenda,* University of British Columbia, Institute of International Relations, Working Paper No. 30 (Vancouver: Institute of International Relations, April 2000); reprinted by permission of the Institute of International Relations, University of British Columbia. Parts of chapters 3 and 5 previously appeared as "The Environment as Geopolitical Threat: Reading Robert Kaplan's 'Coming Anarchy,'" *Ecumene* 3, no. 4 (1996): 472–96; reprinted with permission from Arnold Publishers. Parts of chapters 3, 5, and 8 were published as "Environmental Insecurity: Nature as Geopolitical Threat," in *Of Fears and Foes: Complex Interactive Dimensions of Insecurity in an Evolving Global Political Economy,* edited by Jose V. Ciprut (New York: Praeger, 2000), 79–98; reprinted with permission from Greenwood Publishing Group. Chapter 4 was published as "Geopolitics and Ecology: Rethinking the Contexts of Environmental Security," in *Environment and Security: Discourses and Practices,* edited by Miriam Lowi and Brian Shaw (London: Macmillan, 2000), 84–100; reprinted by permission of Palgrave Publishers Ltd. Chapter 7 and part of chapter 8 were published as "Ecological Metaphors of Security: World Politics in the Biosphere," *Alternatives: Social Transformation and Humane Governance* 23, no. 3 (1998): 291–319; copyright 1998 by Lynne Rienner Publishers; used with permission of the publisher.

Published by the University of Minnesota Press
111 Third Avenue South, Suite 290
Minneapolis, MN 55401-2520
http://www.upress.umn.edu

Printed in the United States of America on acid-free paper

Library of Congress Cataloging-in-Publication Data

Dalby, Simon.
    Environmental security / Simon Dalby.
        p.   cm. — (Borderlines ; 20)
    Includes bibliographical references and index.
    ISBN 978-0-8166-4025-6 (HC : alk. paper)—ISBN 978-0-8166-4026-3 (PB : alk. paper)
        1. Environmental policy—International cooperation.   2. Security, international—Environmental aspects.   3. Environmental protection—International cooperation.
    I. Title.   II. Borderlines (Minneapolis, Minn.) ; v. 20.
    GE170 .D35 2002
    363.7'0526—dc21

                                                                            2002001770

The University of Minnesota is an equal-opportunity educator and employer.

15 14 13 12 11 10 09 08   10 9 8 7 6 5 4 3 2

In fond memory of my mother, Nina Dalby

BORDERLINES

Contents

Preface

This book questions the identities supposedly rendered insecure in contemporary discourses of security as well as the geopolitical assumptions that structure these articulations of danger. It crosses many academic borders and draws connections between places and modes of analysis that rarely intrude on the scholarly discussion of security. The text does so because these intrusions and connections suggest many important things about the reasoning that links world politics to matters of endangerment and threat. Specifically, the following pages examine the incorporation of environmental themes within academic and policy discussions of security in the 1990s. As such this is a book that explores environment and security, but it is also a book that probes the conceptual structures of both to extend the scope of what is now loosely called "critical security studies." Much of the theoretical explanation for this approach will become clear as the argument unfolds in the pages that follow, but the inspiration for the volume comes from more practical matters.

The book had its specific genesis in circumstances far from where the weighty matters of geopolitics or global security are usually studied and discussed. This book is in part my response to a number of questions posed at a meeting of the Canadian First Nations Environmental Network held in June 1997. The meeting occurred in a building on the grounds of Mi'kmaq activist Sulian Herney's house in the Eskasoni reserve on Unama'ki, better known to English speakers as

Cape Breton, the island that constitutes the northern part of the Canadian province of Nova Scotia. Sitting in a circle of activists, listening to them recount the struggles and the devastation of various peoples and landscapes across Canada and, in one case, in the United States, I was struck by the inadequacy of the term "environment" to describe these concerns. Indeed, a number of activists very obviously stumbled in their accounts of their peoples' situation when they came to using the term itself. Alert to the political difficulties of discourse I began to wonder yet again about the term and about why such an obvious part of contemporary vocabulary was not fitting well into the stories native activists were telling. The term "security" wasn't used at all, but insecure these people certainly were, in very many senses.

Thus the book started from empirical research on an apparently very different topic that turned out to be another way of asking questions about environmental security. These concerns then crystallized in a place far, in many senses, from the realms where international politics is practiced—far, too, from the university library, seminar room, or the professorial study. I began to wonder also what the whole scholarly and policy discourse on global environmental politics and, more specifically, environmental security, to which I was a contributor, might possibly have to say to these activists. Surely if it had any merit it must have things to say to these people who were suffering numerous problems of which environmental devastation was an important component. It struck me that it was precisely such people, whose environments were being degraded by various processes, who were often portrayed as a potential security threat in the literature on environmental security. The concerns of scholars interested in the relationships of environment and conflict seemed very distant indeed from the lived realities of the participants at this meeting. The apparent distance from these inspirations on the lakeshore of the Bras d'Or to the final text of this book is, however, key to understanding the topic and the approach to the whole volume.

While in Eskasoni I also began to ponder the links between environmental security and the growing focus in the discipline of geography on political ecology. This critical synthesis of scholarship in development, ecology, hazards, and economics seemed largely absent from the environmental security discussions although it should have had much to offer. Geographers have also dealt with matters of aboriginal rights and land title and the questions of the legacy of imperialism and colonization, matters that are often glossed over by developmentalist assumptions, if mentioned at all, in the environmental security literature. The converse is also true; until recently only a few geographers had engaged either the public debates or scholarly literature on environmental security.

This book attempts to bridge some of the conceptual gaps between these discourses and tangentially raises the question of how scholarship might be useful to activists in apparently remote places, the "people in the way" of numerous developments. But more important, taking inspiration from Michel Foucault, among many others, it takes the critical task seriously by asking questions about how questions are formulated. It probes into the assumptions about environmental security and global discussions of politics. It also tries to think through how politics and scholarship might be rethought if the perspectives of the native activists in places like Eskasoni are worked into the analysis.

The perceived refusal on the part of government bureaucrats to think hard about native activist concerns has been one of the irritants in aboriginal politics in many places. The problem has not been just their unwillingness to do so; more serious, even the possibility of doing so is frequently foreclosed by the technical and administrative arrangements and concepts that structure modern government. The blindspots in contemporary thinking are often replicated in attempts to extend a managerial mentality to the complex political dimensions of security. This problem is acute where environment is concerned because the imperial roots of conservation and resources management thinking and practice are an important, if frequently forgotten, dimension of the contemporary crisis. Focusing on the aboriginal perspectives allows some of this amnesia to be tackled in a way that links global concerns with very local consequences; more important, it demands a critical reexamination of the taken-for-granted premises in contemporary discussions of environmental security.

In February 1998, I found myself a very long way from Eskasoni, in rural South Africa, an invited guest on the Spioenkop game reserve in the foothills of the Drakensberg Mountains. I had gone to South Africa for family reasons and for a vacation to escape the rigors of winter in Ottawa, but the extraordinary circumstances of this beautiful place forced me once again to confront the themes that I have subsequently developed in this book.

Driving through the rolling hills and the rich farmland of KwaZulu Natal, one could nearly have been in parts of the Midwest in the United States. But once I left these lowlands and drove toward the mountains, the small plots on the rapidly eroding hillsides farmed by the "African" population presented a stark contrast to the affluent industrial farms I had passed a few miles earlier. The full meaning of the local geographical parlance referring to "the Africans' area" became very apparent. I wondered what the activists in Eskasoni might have made of the dispossession of these people by the "European" industrial farms interspersed with the forestry plantations on the lower slopes. I also wondered how Thomas Homer-Dixon's analysis of environmental scarcity, partly researched some years earlier in South Africa by a former graduate student of mine, Valerie Percival, played out in this particular village. Above all else I reflected on the complex geopolitical and orientalizing cultural codes that specified the identities of people in this particular part of KwaZulu Natal as "European" and "African."

Walking in the reserve, I noted the sites where some Africans' huts had formerly stood. Once again I was presented with the practice of removing residents to create parkland, of limiting human occupation to tourists, some select staff, and European residents around the administrative hub of the reserve. This was very much an artificial creation of grassland for the "endangered" herbivores built around a reservoir designed to provide water for the distant residents of Johannesburg. My boat trip a few days later on the lake was interrupted when the pilot stopped to give some instructions to a group of women on which species they were to chop down in one part of the reserve lands. Park boundaries here, as elsewhere, do not prevent the migration of unwanted species.

The reserve administrators were trying to maintain a particular mix of species in the reserve and trying to provide local employment that would be useful in itself but might also gain the support of the local population for the continued operation of the reserve. I wondered if one of the patiently waiting women might once have lived in a hut on the reserve, and I pondered the economics of selling the "right" to hunt to American hunters as a way of providing the cash to pay locals for clearing the brush. This proposal was being debated as a possibility due to the lack of a top predator in the artificial ecosystem of the reserve and the need to generate additional cash for reserves in postapartheid South Africa. Considering what kind of masculinity might be reinforced by shooting such a "trophy" and what such a sensibility might mean for notions of environment forced me to reflect on why I found this Edenic spot, with its grasses, thorn trees, and animals so charismatic, on the European orientalizations of "exotic" lands, and on the geopolitical implications of such specifications in my own cultures. Discussions about rising crime rates, carjackings, and Europeans' anxieties about their personal safety reminded me of Robert Kaplan's alarmist prose about Africa as the model of the political future everywhere.

Having won a decidedly foolish staring match with a rhinoceros, my trip to the top of Spioenkop hill was especially interesting. The rather tired memorials to the troops that died during a particularly bloody Boer War encounter a century earlier reminded the visitor of the violence of conquest and the contested nature of this place. The crucial connection between geography and military activity was obvious in the memorials; a surprise attack had gone disastrously wrong when an officer misread the topography on the top of the hill at night. I wondered how Liverpool Football Club in England had performed that weekend and how many of the soccer fans in the "Kop" end of their Anfield stadium had ever been to this hill to see where so many of their antecedents had died. Apparently because the terraces in Liverpool Stadium looked similar to the topography of the hill, they had been renamed in memorial to the dead imperial troops from Liverpool. All these connections to distant places were in this one unlikely battlefield turned game reserve: warfare and conservation, environment and security. I pondered why they might seem to be unlikely connections as, now safely out of reach of the rhinoceros, sipping my Castle lager, I watched the South African television advertisements for corporate-sponsored sports training programs, reinventing postapartheid masculine identities.

Eighteen months later, coming back to Vancouver on sabbatical to a city where I had been a graduate student and taught for some years prior to moving to Ottawa, I ended up in a third part of the former British empire, one that celebrates its heritage in the province so very aptly named British Columbia. Here aboriginal issues are now regular headline news, the debate over the Nishga treaty a political priority. But in this very obviously British city on the Pacific Ocean, where the nouveaux riches, dot-com millionaires are buying luxury homes in the "English properties," the rapid social changes as a result of the influx of Asian immigration are unavoidable, and so are practical attempts to deal with some pressing contemporary environmental issues. Bicycle lanes have appeared in the city. Forestry practices are under review all over the province. Public transit is a priority in a city that has managed to avoid the ravages of urban freeways. Urban life is being reinvented downtown with huge, new high-rise developments revitalizing the streetside cafés and waterfront parks while reducing commuter traffic from the suburbs. In the city, enterprising organic grocery delivery services are thriving and changing consumers' eating practices as well as their awareness of the environmental dimensions of agriculture. I asked, how does all this relate to environmental security?

This book is an attempt to grapple with all these questions because it seems that putting them all together is necessary if the difficulties that appear under the sign "environmental security" are to be tackled with the appropriate degree of conceptual self-reflection. As such, and in keeping with the ecological motif of many of the arguments in what follows, the text crosses many artificial disciplinary borders; indeed, it argues that borders are in many ways the problem in thinking about environmental security. Just as the African women's labor is needed to preserve the species mix in Spioenkop, so are the borders of various disciplines patrolled and ideological weeds removed or at least contained. But in the process the possibilities of new growth, ecological synergies, and innovative habitat as a result of such colonizations are lost.

It is a rich, but apposite, irony that the geographical discipline in which I was trained, and whose political branch has been so preoccupied with borderlines, has of late been the context for all sorts of interdisciplinary migrations that have enriched the diversity of its enterprise enormously. To borrow the prevalent disciplinary metaphor in these matters, in moving from a preoccupation with cultural roots in parts of space to the recognition of the importance of multiple routes to understanding contemporary places, the complexity of ecological politics has been implied in investigations of many contested identities. This book tries to explicate these themes in a way that offers a contribution both to political geography and to security studies while unsettling both by its frequent border transgressions.

I have incurred many debts along the way. My thanks first of all to Sulian Herney and the activists of the First Nations Environmental

Network for allowing me to attend their meetings in Nova Scotia in June 1997. My thanks also to my aunt, Marcia McGillivray, for her invitation to Spioenkop and to the staff there for making me so welcome. Although the experiences in Eskasoni and Spioenkop crystallized my thinking into what has become this volume, I have been working on the literature linking environment and security since 1990. Through the decade I have incurred many intellectual debts, not all of which I can acknowledge here, not least because so many conversations, not to mention e-mail messages, over the years have shaped my thinking in many ways. Most importantly, my partner in intellectual subversions, Cara Stewart, deserves my thanks for endless hours of discussion and much patience with my authorly preoccupations. Her parallel concerns with the limitations of conventional political thinking have been a rich source of inspiration despite the fact that crucial papers and books have repeatedly migrated to her study at the wrong moment.

Ongoing conversations with Thomas Homer-Dixon, Matthias Finger, Geoffrey Dabelko, Richard Matthew, Betsy Hartmann, and Daniel Deudney have enriched my thinking about environmental security; likewise, Gearóid Ó Tuathail, Klaus Dodds, Matt Sparke, Jo Sharp, and John Agnew have been especially helpful in my thinking about critical geopolitics. R. B. J. Walker, David Campbell, Bradley Klein, Jim George, Michael Shapiro, Michael Williams, Keith Krause, Jennifer Milliken, Ken Booth, and many other political scientists have helpfully engaged my ideas about politics and security over the years. Bruce Braun, Derek Gregory, and the other participants in the 1999–2000 University of British Columbia Green College lecture series on "Nature, Culture, and Colonialism" were especially useful stimulation while writing this book.

Thanks to Celeste Wincapaw for research assistance in the preparation of the original paper on Robert Kaplan, presented as "Neo-Malthusianism in Contemporary Geopolitical Discourse" at the annual meeting of the International Studies Association, Chicago, February 1995. The bulk of this appeared in *Ecumene* in 1996 and is reworked here in chapter 2 and some of chapter 5. Early versions of some parts of chapters 3, 5, and 8 were presented as "Lacunae and Lapses: Reading the Silences in Environmental Security Discourse" at the annual meeting of the International Studies Association, Toronto, March 1997. My thanks to Jose Ciprut for his editorial work on the version that appeared in his edited volume *Of Fears and Foes,* published by Praeger in 2000.

Chapter 7 and some of chapter 8 were presented at the British International Studies Association annual conference in Leeds in December 1997. My thanks to R. B. J. Walker for his contribution in preparing that paper for publication in *Alternatives* in 1998. A version of chapter 4 was presented as "Environmental Geopolitics" at the annual meeting of the International Studies Association in Minneapolis in March 1998. My thanks to Miriam Lowi for her editorial suggestions on the version that appeared in her coedited volume *Environment and Security: Discourses and Practices,* published by Macmillan in 2000. Both these papers were prepared with financial support from a Carleton University research achievement award for work on "A Critical Geopolitics of Environmental Security."

Most of chapters 1 and 8 was presented as a seminar to the Strategy and Security Research Group, University of Aberystwyth, 15 September 1999, and to the University of British Columbia Institute of International Relations, 28 October 1999, and published as *Geopolitical Change and Contemporary Security Studies: Contextualizing the Human Security Agenda,* Working Paper No. 30 (Vancouver: University of British Columbia, Institute of International Relations, April 2000).

In addition to the students in my "Advanced Political Geography" seminar course at Carleton University, who have struggled with the themes in this book in classroom meetings, readings, and papers, I have been fortunate to work with many enthusiastic graduate students whose interests have connected directly to the environmental security problematique. They have influenced my thinking in many ways. In particular, my thanks to Denis Arsenault for work on hazards; Steve Gwynne-Vaughan on development and security in Africa; Valerie Percival on conflict in South Africa and Rwanda; Anastasia Chyz on environmental degradation and Haiti; Mathew Coleman for showing me that what I was doing was a critique of Lockean governmentality; and to Shona Leybourne for ideas on food security, Africa, and many related matters. William Hipwell has taught me much about aboriginal and environmental politics and a little about science fiction. He was also an excellent researcher and interviewer for the project on "Community, Identity, and Environmental Threat," which is drawn on in chapter 5. Thanks also to Chad Briggs for making me think seriously about Mars, Mary Hutcheon for regimes and biodiversity, and Dan Riesborough for discussions of ecology and ontology.

My colleagues at Carleton have helped, too, in providing me with opportunities to present my ideas in a number of fora and providing informal advice and sympathetic ears on many occasions. Thanks to Susan Tudin for endless help in the library; David Long for international relations thinking; Iain Wallace and Nancy Doubleday for ecology and development; Fiona Mackenzie and Madeleine Dion Stout, collaborators on the "Community, Identity, and Environmental Threat" work on Cape Breton, as well as advisers on aboriginal matters, feminism, political ecology, Africa, and development; Mike Brklacich and Ken Torrance for environmental security discussions; and Mike Smith and Chris Burn for input on climate change and related concerns.

In the early 1990s a Barton fellowship from the now, alas, defunct Canadian Institute for International Peace and Security was invaluable in starting my research into environment and security. The field research in Cape Breton was supported by a research grant from the Canadian Social Sciences and Humanities Research Council. The Institute of International Relations at the University of British Columbia provided the ideal institutional context to write the first complete draft of the manuscript. My thanks to Brian Job, director of the institute, for his hospitality and support, and to Will Bain for lengthy conversations about all manner of political things. Geoffrey Dabelko and the staff of the Woodrow Wilson Center Project on Environmental Change and Security have been most helpful over the years, not least in stimulating me to critical reflection by requesting that I review some key works in the field.

I am grateful to Jon Barnett, Mathew Coleman, William Hipwell, and Matthew Paterson, who each provided a detailed critical reading and extensive helpful comments on the first draft of the complete book. Given the numerous heresies that appear in the following pages, it is especially important to note that none of the people or institutions mentioned here bears any responsibility for the contents of this book. My follies and foibles are mine alone.

Introduction
Environment, Security, and Geopolitical Discourse

For security, the genealogist would insist, is not a fact of nature but a fact of civilization. It is not a noun that names something, it is a principle of formation that does things. It is neither an ontological predicate of being, nor an objective need, but the progenitor instead of a proliferating array of discourses of danger within whose brutal and brutalizing networks of power-knowledge modern human being is increasingly ensnared and, ironically, radically endangered.

MICHAEL DILLON, *POLITICS OF SECURITY*

ENVIRONMENT AND THE NEW SECURITY AGENDA

Since the end of the cold war, environmental concerns have been an important part of the discussions of global security in North America and Europe. Whether under the rubric of environmental security, as part of a broad human security agenda, or one among a number of new global dangers, environmental themes are now part of the calculus of international politics and part of the scholarly debates in international relations. The potential for warfare over access to scarce resources and intact environments looms over these discussions and has triggered a considerable amount of scholarly research, not a few books, and much public discussion. In light of concerns about ozone holes, climate change, biodiversity, and related matters, the traditional geopolitical themes of great power rivalries, access to resource supplies, and governance at the largest scale are now extended to encompass environmental themes in the policy-making institutions in Washington, Brussels, and elsewhere.

The broadened agenda of security in the post–cold war era has included numerous other nontraditional threats including drugs, disease, weapons proliferation, failed states, demographic changes, and a host of other items.[1] But the single theme in this broadened agenda that garnered both the greatest attention and produced the most intense conceptual and political debates in the 1990s was the environment. Despite more recent scholarly attention to humanitarian interventions, conflict termination strategies, and related matters, the environmental theme remains important. While the international security agenda has changed through the past few decades, exactly how this new-found concern with environment should be related to security is still much less than agreed upon despite considerable discussion leading up to and following the 1992 United Nations Conference on Environment and Development (UNCED). The 2002 World Summit on Sustainable Development continues the debate into the new millennium.[2]

Through the 1990s the emphasis on environment focused critical scholarly, and some policy-making, attention on the pressing need for adequate conceptualizations of both environment and security in trying to research the supposed connections between the two. The more critical early contributions challenged the conceptual frameworks used in both the scholarly research and policy discussion. Especially, they suggested, when related to the concern with environment, broadening the themes addressed in a security framework suggests a series of complex dilemmas and anything but a simple framework for evaluating policy options. By analyzing this debate, and especially by probing in more detail into its ontological premises, we can learn much, not only about the literature on environmental security, but more generally about the larger policy and scholarly discussions that invoke the term "security."

The debate about definitions of security and the identification of environmental sources of political threat is part of the post–cold war scholarly discussion of the appropriate concepts and methods for the study of international security.[3] Adding such items as diseases, migration, economic dislocations, and the gamut of human rights issues to the security agenda also suggests, crucially, that new tools and ways of thinking are needed to offer an all-encompassing approach to multitudinous threats. In 1995 Davis Bobrow suggested in his presidential address to the International Studies Association that a medical metaphor—depicting security analysts as finding, diagnosing, and prescribing remedies to all sorts of ailments in the body politic—might best convey what practitioners of security now do as part of their normal professional activities.[4] The need for a new metaphor comes because it is apparently necessary to broaden the ambit of security to deal with the geopolitical difficulties of the post–cold war era.[5]

But new metaphors, and specifically new professional identities, require more than the addition of new items to the established agenda; they require a larger intellectual debate about the scope and purposes of the security enterprise. In addition, as the epigraph to this chapter suggests, because invoking security is a political act and the discourses that construct dangers and endangered subjects are far from natural or neutral reflections of an independent reality, the larger social and political contexts within which such discourses are invoked should also be given analytical attention.[6]

CRITICAL SECURITY STUDIES

According to Ole Wæver, just such a "fourth debate" is either under way or recently concluded in the discipline of international relations.[7] Supposedly following from earlier debates about idealism or realism, historical or scientific method, interdependence or anarchy, this fourth debate introduced more powerful critiques of the discipline that suggest that the whole enterprise of knowing the world and discussing politics in its terms is part of the problem that needs to be addressed. More specifically, the subfield of security studies has been challenged by a number of theoretical innovations that call into question the ontological givens of the cold war discourse. Drawing on disciplinary perspectives from outside international relations and, more broadly, on the debates in social theory about method and the politics of knowledge, these studies have challenged the identities that international relations is premised upon and the construction of the communities that have thus been understood as endangered.

Security is a derivative concept, one that already assumes that there is something else, outside of itself, that has to be secured. This is the obvious starting point of such a critique. It is where Keith Krause and Michael Williams start their introduction to the volume that explicitly suggested the need for a scholarly enterprise called critical security studies.[8] What is to be secured, how, and by what means are questions conventionally answered by focusing on the state. But when environment is added into the consideration of security, it may well be that the state is no longer the obvious referent object of security. The crisis of sovereignty, so often linked to states and security, in the face of globalization is especially clear when environmental themes are added into the discussion. More so than most topics, the environmental theme in critical studies makes the fact that security is a political construction in specific contexts unavoidable. But as this book also demonstrates, the constitution of context itself is crucial to the invocation of endangerment.

The other crucial point about critical security studies is the attempt to link critical thinking about security to some form of a politics of emancipation. Clearly spelled out by Ken Booth in the early 1990s and more recently addressed in detail by Richard Wyn-Jones, the suggestion is that security studies has long been implicated in the practices of violence that have rendered people vulnerable to many dangers, most obviously to the dangers of nuclear war and massive state violence.[9] Discussed in terms of the politics of security dilemmas and the violence perpetuated by strategic modes of reasoning, emancipation suggests freedom from such structural constraints to human aspiration. In the terms of Mick Dillon's epigraph that opens this chapter, security is something that, ironically, endangers. Quite who or what political identity is being secured is an explicit theme of the critical studies literature because, by its geopolitical practices, security is such a powerful producer of endangered identities. But, as this book implies in later chapters, engaging with environmental themes and linking them to security makes the question of emancipation especially difficult to conceptualize because the constitution of the modern urban emancipated subject itself is challenged by critically rethinking human ecologies.

In thinking along these lines, inevitable questions of how to study such things arise, and security scholars have recently ventured into the fields of sociology and anthropology to extend the conventional analyses.[10] The sociological turn has enhanced the focus on culture and identity but all too frequently remained caught in the strictures of state-based thinking where states, it is reasserted, are the only entities that have identities and cultures that matter.[11] Some of the lit-

erature in anthropology, in contrast, and consistent with the historical division of academic labor where anthropology studied peoples
without states, escapes this assumption and opens up a larger canvas
of consideration of who is insecure and the sources of their insecurity.[12] Here security studies also collides with feminist analyses of
militarism and the international politics of gender, violence, and pa-
triarchy. In particular, Cynthia Enloe's empirical disruptions of the
conventional codes of gendered diplomacy and militarism have long
challenged the logics of international relations, while Carol Cohn's
feminist anthropology of nuclear strategy long ago rendered its claims
to objectivity and scientific authority untenable.[13] These interventions place the political questions of security explicitly back on the
agenda of discussion.

These critiques pose fundamental questions about the identities
being rendered secure by contemporary practices of violence and by
the geopolitical divisions of political space into states, blocs, and geographically bounded political communities. More specifically, the environmental dimensions of the security debate, often understood as
a challenge to sovereignty where pollution and such factors as climate change transcend state boundaries, unavoidably raise the political questions of secure insides and external threats.[14] Most explicitly,
the theme of environment, which is by definition that which surrounds
an entity, directly raises complex questions about modern spatial
practices of politics as soon as the obviousness of the inside of the
modern identity is confronted by the possibility of such an external
threat.

As the rest of this volume reiterates, once the natural environment
is taken seriously as a security theme, the conventional assumptions
about states as the containers of political community can no longer be
taken for granted. This in turn demands a reconsideration of the modern identity that is supposedly endangered by the environment. Here
the argument adds to the thrust of a growing number of critical interventions that examine how conventional reasoning practices and
assumptions preclude engaging with the possibilities of world politics.[15] The crucial point for much of what follows in this book is that
the ontological limitations of international relations thinking are especially acute when matters of global environmental politics and environmental security are addressed.

CRITICAL GEOPOLITICS

Matters of the spatial specifications of global politics, violence, and identity are now also discussed under the label of critical geopolitics.[16] This literature, which overlaps with critical international relations thinking, has suggested that the taken-for-granted geographical categories and descriptions of the earth (literally "geographs") in political thinking are an important matter for critical analysis.[17] Geopolitical reasoning, that is, using geographical categories as part of the practices of representation among foreign policy makers and politicians, specifies the world in particular ways that have political effects.[18] Crucial to all this is the reduction of complex geographical realities to simple strategic entities to facilitate discussion and political action. In the process, contested social constructions become accepted terms for discussion with powerful political consequences.[19] This in turn is related to larger questions of the production of geographical knowledge and how specific forms of knowledge facilitate state practices.[20] It is also important to emphasize that the general cultural geography of world order, as well as the geopolitical codes of policy makers, is routinely reproduced in the quotidian practices of social life.[21]

Through time the importance of particular places change. Even their constitution as places, or more specifically as geopolitical entities, changes. Regions are eliminated from geographical reasoning as cultures, boundaries, understandings, and technologies change. This crucial recognition was an important part of the new geopolitical scholarship of the 1990s that rejected the earlier formulations of geopolitics with its determinist themes and assumptions of unchanging geographical verities.[22] The political significance of this recognition is considerable. For example, this is especially clear in Klaus Dodds's examination of the shifting geopolitical evaluation of Antarctica through the twentieth century.[23] At various times Antarctica was a token of geopolitical rivalry. During the cold war it was primarily a concern involved in various national security scares. More recently it has been a matter of various treaty negotiations concerning natural resources in its oceans and the preservation of its landscape against the damage of mining proposals. Now it has become a place of international collaborative scientific research monitoring climate change and ozone depletion, as well as an exotic destination for tourists. Antarctica it still is, but its political significance and the knowledge practices used to describe it and to construct appropriate governance institutions have changed dramatically.[24]

How expert specifications become part of the rationalizations for state and corporate activity, or oppositional politics, is now an unavoidable part of any discussion of environmental or security politics.[25] The point is not simply that knowledge is power, but that knowledge and power are imbricated in each other in complex discursive formations of what Michel Foucault sometimes called power/knowledge. In the terms of contemporary poststructural thinking, political objects are thus best understood as discourses: systematic modes of the specification of objects related to procedures for designating, studying, and disciplining such entities. They are part of both the social practices of everyday life and international politics. The best way into these matters is often through an explicitly ontological investigation. The most important ontological categories in political discourse are frequently the simple, taken-for-granted geographs or specifications of the world, sometimes understood in other geographical writing as "metageographies" or the "spatial structures through which people order their knowledge of the world."[26]

It is often useful to draw loose distinctions between the realms of formal geopolitics (where intellectuals of statecraft in the academy and think tanks formulate the themes of foreign policy making and international relations), practical geopolitics (used by politicians and policy makers in the articulation of the practices of statecraft), and the more encompassing routine cultural productions of places and geopolitical identities in the mass media of "popular geopolitics."[27] Critiques of conventional assumptions are often effectively anti–geopolitical reasoning insofar as they challenge the ontological specifications of politics and state claims to identity. As chapter 1 will elucidate, these forms of discourse overlap, especially in claims to endangerment, when official discourses try to render particular processes as security threats to a particular identity.

Here the critical geopolitics literature links directly to contemporary thinking in international relations, and particularly to such formulations as David Campbell's use of the theme of ontopology to link ontology with topos, being with specific places.[28] As Campbell's analysis of the violence in Bosnia clearly shows, the cartographic specifications of reality that can be invoked to justify claims to identity and territory are powerful modes of reasoning with, in this case, violent consequences. But similar ontopological claims now work on the global scale; in the tropes of global security it is the earth itself that is an endangered place, requiring, in some formulations, nationalist-style rhetoric of salvation.[29] The spatial endangerments of security discourse are linked to the understanding of this place as the planet itself, although the exact source of the danger is frequently not clear.[30]

The spatial disaggregation of the earth's surface into multiple territories, administrative regions, parks, development zones, and private property is also part of the contemporary practices of state administration and the codes of rule. But as Neil Smith has argued, these "productions" of space are also "productions" of nature.[31] By dividing up territories and administering specific places according to codes of management, nature is turned into an environment containing various resources. Land "clearing" is, of course, an environmental change as it turns forest into farmland. Linked to the colonial practices of enclosure and land "improvement" in the processes of the expansion of European power in the past half millennium, the spatial disaggregation of modernity has involved a dramatic series of ecological changes as it has repeatedly remade maps of power and identity. In Michael Shapiro's apt phrasing, what is forgotten in the naturalizations of geographical categories is that cartography is often a violent practice.[32] But now it is necessary to link these matters of critical geopolitics with matters of critical ecopolitics to understand contemporary environmental dangers as being, in part, about the political boundaries that produce both nature and space.[33]

THE BOOK

Broadening the ambit of security inevitably requires that larger questions be posed than those traditionally asked in security studies. As such the book asks cross-disciplinary questions precisely because a crucial part of the argument is that the limits of traditional disciplinary approaches to security, literally how security has been disciplined, are part of what needs to be investigated. To do so requires refusing the key assumptions in international relations scholarship as to the primary importance of the category of the state and the location of political community within states. As many writers have recently suggested, the question of the global environment challenges claims to sovereignty. Likewise, it challenges the assumptions implicit in the formulations of security. Most important of all, however, it challenges the identities that are endangered by contemporary processes. The book is thus a critique of geopolitical reasoning, but also of the taken-for-granted assumptions in environmentalist discourses. It draws on the discussions of environmental history and some contemporary research in ecology to challenge contemporary use of the terms "security" and "environment."

This book makes no promises of a better environmental security, a more correct methodology, right answers, or improved policy. What it does do, in parallel with contemporary critical thinking in a number of disciplines, is unpack the implications of the ontological categories of geopolitical and security discourse. If the world does face the many ecological disruptions that are the cause of discourses of environmental security in the first place, then scholars should critically examine the commonsense assumptions that so often shape environmental thinking, activism, and security discourse. Likewise, if the dominant political and economic systems of the twentieth century, which security was largely about maintaining, contributed to, or oversaw the destruction of, so many natural attributes of the planet, it seems necessary to investigate these discursive practices, too, and challenge the obviousness of their specifications of danger.

To make this case, the argument is presented as an interconnected cumulative narrative where each chapter draws on themes developed in the previous ones. Chapter 1 is a detailed discussion of the extension and reformulation of the concept of security in the field of international relations and beyond in the 1990s. Part of the argument is about the extension of matters portrayed as threats requiring a security response. In particular, environmental themes have been added into discussions of international security. The questions of identity and who specifies danger are crucial to this discussion. Chapter 2 deals directly with the high-profile public articulation of the themes of environmental security in the 1990s, which appeared in Robert Kaplan's article "The Coming Anarchy," published in the *Atlantic Monthly* in 1994.[34] Unpacking Kaplan's geopolitical premises raises important questions about how the larger discourse is constructed. The findings of the subsequent scholarly research on environment and conflict, discussed in chapter 3, have some fairly clear conclusions but are often at odds with both the conventional geopolitical formulations of security and Kaplan's worst fears.

Chapter 4 directly challenges the geopolitical premises of the discussions of environmental security, suggesting that its historical amnesia needs to be corrected in light of the findings in chapter 3 and recent research in environmental history. Chapter 5 focuses on the international connections that chapter 4 shows are often missing in conventional treatments of environmental regimes and security. It also draws some brief but important insights from aboriginal politics, matters not usually considered to be part of the discussions of environmental security. Focusing on the contemporary predicament of native peoples in North America links history and environmental changes with dispossession and insecurity. Chapter 6 then looks at alternative ways of conceptualizing contemporary environmental degradation and asks, in light of the inadequacies of the geographical specification of the issues of environmental security discussed in chapters 4 and 5, how we might more adequately reconsider the geopolitical precepts of both environmental change and security.

Chapter 7 extends this by drawing on ecological theory to offer a contrast with conventional understandings of security and so further revealing the limitations of the assumptions as to what is being secured. In doing so, it suggests a number of new formulations that might provide a useful heuristic template to reevaluate the assumptions of international relations thinking in general, and security thinking in particular. Chapter 8 briefly returns to the concerns about rethinking security and the possibilities of recrafting security studies and policy making in a more critical manner, sensitive to the ecological reasoning outlined in the latter half of the book. In light of these alternative formulations, chapter 9 thinks about the future and asks what precisely is so important that it must be secured. There are no blueprints here or any detailed reiteration of environmentalist policy proposals, just some suggestions about alternative ways of thinking that might at least usefully clarify the politics of some pressing contemporary dilemmas.

There are a number of closely complementary perspectives and academic debates that this book might have drawn on, or contributed to, more directly. For example, the discussions of world systems theory, imperialism, and dependency, as well as the larger conceptualizations of international political economy, have a considerable amount to say about the themes in this volume. But in the interests of clarity, brevity, and focus, I have avoided a detailed engagement with this literature.[35] The extensive literature on global governance, environmental regimes, and related themes is likewise obviously important, but not directly related to many of the themes that link security to environment once its assumptions are critiqued as this book does in chapters 4 and 5.[36] Nor, apart from the discussion at the beginning of chapter 1, does this book explicitly engage with the broad themes of contemporary world order and the changing nature of global politics, although there are implicit connections with these themes in most of what follows, especially in chapter 9.[37]

This book might be read as an explication of the themes of eco-imperialism in the World Order Models Project literature and in Richard Falk's recent writings.[38] A general sense of empire as an overarching alternative narrative for understanding contemporary transformations is implicit in the whole book, although this was all written prior to the publication of Michael Hardt and Antonio Negri's explicit theorization of the theme, which ignores most of the environmental dimension.[39] Much of the narrative that follows, and the theme of fear and aspiration in chapter 2 in particular, might be written in a psychoanalytical register of desire. However, the theme of identity and the critique of security as a metaphysics of domination and geopolitical containment, which is implicit in all that follows, offers a more appropriate interpretive schema to link environment with security. While chapter 8 directly raises issues about the future of security studies, I have eschewed detailed engagement with the contemporary debate about "positivism," "constructivism," and "poststructuralism" in the discipline of international relations.[40]

METHOD, CRITIQUE, ENVIRONMENT

That said, however, I have been heavily influenced by many "poststructuralist" writers, and although the sources, arguments, and evidence used in what follows are much less than obviously "post," the themes of space, identity, and colonization, and my overall strategy of problematicizing the taken for granted, fit with its ethos. This book is a work of criticism, a contribution to investigations in a number of overlapping academic fields, as well as an argument with the ongoing discussions about security in North America and Europe. As such I follow David Campbell's reminder to readers of his *Writing Security,* regarding what Michel Foucault said about critique:

A critique is not a matter of saying that things are not right as they
are. It is a matter of pointing out on what kinds of assumptions, what
kinds of familiar, unchallenged, unconsidered modes of thought the
practices that we accept rest.

We must free ourselves from the sacralization of the social as the
only reality and stop regarding as superfluous something so essential
in human life and in human relations as thought. . . .

Criticism is a matter of flushing out that thought and trying to
change it: to show that things are not as self-evident as one believed,
to see that what is accepted as self-evident will no longer be accepted
as such. Practicing criticism is a matter of making facile gestures
difficult.[41]

While I do not wish to suggest that the environmental security discourse is facile, in many ways its premises, and its assumptions about
"environment" in particular, are not nearly as self-evident as its many
authors sometimes apparently think. And how we, of whatever fictional community, think leads not only to how we act politically, but
also to our understandings of who we are, what we value, and what
we are prepared to countenance to protect our self-preferred identities. This is the very stuff of security.[42]

This book does not engage in any detail with either the history of
environmental philosophy or the specific compatibilities of various
streams of environmentalist writing and politics with international relations or security studies. Such efforts have been undertaken recently by other scholars.[43] Neither does it revisit the major debates in the
1970s about the limits to growth, steady states, and political alternatives that were, in some ways, precursors to the contemporary discussions.[44] Invoking various narratives of environment, and in particular a critique of colonizing practices, does not simply suggest that
this environmental story line offers some transcendental or objective
discourse that provides the singular truth from which policy can be
derived. Such themes are very much the stuff of both environmental
and international politics; nature has been invoked in numerous contexts to rationalize many political programs.[45] Rather, the counter-
narrative that follows aims to disrupt the conventional formulations
of "environment" as a technical matter for expert regulation or as a
matter for global management, big science, and specifically security
discussions.[46] In the process it will show how the politics of invoking something called "environment" works and will suggest that the

geographical presuppositions in the discourse are especially important. By focusing on historical antecedents of the contemporary crisis, the terms in which we currently understand both environment and security can be criticized and their politics investigated.

This is not a matter of disputing the claims that environmental change is or is not occurring, or challenging the technical practices of numerous disciplines. Whatever the finer points of the specification of "global" ecological processes, there are many reasons for great concern on all manner of issues and in numerous contexts from biodiversity decline to stratospheric ozone holes, and from rising childhood asthma incidence rates to the contemporary sufferings of marginalized peasants and refugees.[47] What is most worrisome to anyone who observes these matters is not any single concern—be it climate change, biodiversity loss, synthetic chemicals, deforestation, long-lived radioisotopes, or any one of many other matters—but the totality of the disruptions caused by modern industrial systems and the consumption of their products, whose cumulative and increasing impact has reached into all parts of the biosphere. This is, of course, both the strength of the environmentalist argument and, given the diversity of its subthemes, simultaneously its greatest political difficulty. The focus in some of what follows in this volume is on the use of fossil fuels, both because they are so integral to contemporary modern modes of economic existence, and hence can be read as symptomatic of the larger condition, and because at a very simple level, by literally turning rocks into air, their widespread use draws attention directly to the anthropogenic alterations of basic planetary systems.

What is most important for the argument that follows is a recognition that contemporary endangerments materialize within political and cultural contexts that constrain, in important ways, how these matters are represented. The political and economic order of modernity is rarely fundamentally questioned in such discussions. The commodification of "nature" is taken for granted as an unavoidable necessity.[48] In particular, despite all the ambiguities of modernity, the developmentalist assumptions that suggest that each state will become modern along approximately similar trajectories of industrialization and modernization are implicit in most conventional analyses.[49] Environmental discourses occur within larger discursive economies where some identities have more value than others, and crucially where the dominant development and security narratives are premised on geopolitical specifications that obscure histories of ecology and resource appropriation. They also frequently operate in discursive modes that reassert geopolitical identities precisely by how they specify other peoples and places.[50]

Environmental politics is very much about the politics of discourse, the presentation of "problems," and of who should deal with the concerns so specified. These discourses frequently turn complex political matters into managerial and technological issues of sustainable development where strategies of "ecological modernization" finesse the questions by promising technical solutions to numerous political difficulties and, in the process, work to co-opt or marginalize fundamental challenges to the contemporary world order.[51] In Tim Luke's apt summation: "Underneath the enchanting green patina, sustainable development is about sustaining development as economically rationalized environment rather than the development of a sustaining ecology."[52] Linking such themes to security, with its practices of specifying threats and its managerial modes for responding to dangers, suggests a broad congruency of discourse and practice.[53] But what ought to be secured frequently remains unexamined, as does the precise nature of what it is that causes contemporary endangerments.

Like other disciplinary endeavors, both environmental management and security studies have their practices for the delimitation of appropriate objects, methods, and procedures. Making these explicit and showing how they both facilitate and simultaneously limit inquiry is an unavoidable task for any study that takes Foucault's formulation of critique seriously.

Challenging conventional wisdom is rarely easy, and disrupting geopolitical categories can be especially "unsettling." Asking unsettling questions about the identities of those who think in the conventional categories is not easy either. But it seems very necessary now, given the limitations of both the security and the environmental discourses we have inherited from the past and the pressing need to think intelligently about what kind of planet we are making.

1

Rethinking Security Studies

The environmental threats facing the planet are not simply the result of sci-entific miscalculation. Nor are they merely the consequence of ill-conceived management decisions. Ironically, it is the notion of security upon which our entire modern worldview is based that has led us to the verge of ecocide. . . . In less than a century the practice of geopolitics has pushed the world to the brink of both nuclear Armageddon and environmental catastrophe, forcing us to reconsider the basic assumptions of security that animate the modern worldview.

JEREMY RIFKIN, *BIOSPHERIC POLITICS*

GEOPOLITICAL CHANGE AND "SECURITY"

As numerous commentators have noted, people in Europe and North America are living in a geopolitical period often simply called the post–cold war. We have been doing so for a decade. Despite the in-creasing popularity of discussions of globalization, it is noteworthy that there is still no widely agreed-upon term for the period and that it is still often defined in terms of a previous era. It suggests a period in which geopolitical identities are in flux and in which there is no hegemonic understanding of the world order and the roles that par-ticular states play. Economic events in Asia in the 1990s have also challenged the more simplistic assumptions of either coming Asian dominance in world affairs or geoeconomic competition as a major concern for security analysts. Persisting conflicts and humanitarian disasters in various parts of the world continually suggest widespread human insecurity. But how to interpret, much less name, the contemporary period remains in thrall to the conceptual apparatus brought to bear on the present. Thinking about now in large terms remains difficult.

The geopolitical options canvassed by North American analysts of the contemporary scene are many. Robert Kaplan's "coming anarchy" of ethnic conflict and crime induced by environmental collapse will be discussed later in this book. Samuel Huntington, in a much cited rearticulation of realist pessimism concerning the inevitability of clashing autonomous entities, proclaims a situation where we are doomed to intercivilizational conflict in perpetuity.[1] Well before his theme was appropriated by James Bond screenplay writers for the aptly titled *The World Is Not Enough,* Zbigniew Brzezinski updated classical geopolitical thinking to focus on central Asia as the key area in the new global "chessboard."[2] Benjamin Barber speculates about a future fate of jihad and McWorld in which ethnic fragmentation coexists with economic globalization.[3] Many writers suggest some form of chaos or at least a world that can be understood best as unruly, precisely because the ways in which it is changing are unclear and much disorder prevails.[4] Richard Barnet and John Cavanagh suggest a future dominated by global corporations in which states are of declining importance in making the major decisions about how populations live.[5]

In focusing on the corporate dimension of the changing order, Barnet and Cavanagh may be closest to the dominant neoliberal assumptions of the world's political elites. Kenichi Ohmae has gone as far as declaring an end to the nation-state.[6] "Globalization" is the term most frequently used now for a period of world order where enhancing global trade and restructuring both states and international regimes to enhance the global reach of international capital is the primary political desideratum.[7] Etel Solingen's recent work, which suggests that states taking a long-term strategy of constructive multilateral economic engagement have better security outcomes in the long run, provides empirical support for the updated modernization theory that apparently underlay the Clinton administration's foreign policy geopolitics of "enlargement."[8] Mary Kaldor's focus on globalization and the geography of contemporary violence suggests that there is an increasingly globalized pattern to the economic system that underlies the violence that frequently occurs in the poorer and more marginal parts.[9]

The lack of clarity concerning exactly what the dominant geopolitical structure is at the beginning of the twenty-first century connects directly to the ongoing debate in "security studies" concerning the subject of security and the meaning of the term in contemporary politics. While national security had some basic geopolitical precepts that were understood widely during the cold war, at least in the Western alliance, national security now is much less clearly understood. In the aftermath of the cold war the larger scholarly and policy debate about these matters has included at least some of the powerful critiques of cold war policies and conceptualizations of national security.[10] All of which has suggested uncertainty about the future and the likely shape of dangers in the future. This sense of uncertainty is also part of the sense of insecurity shaping contemporary political anxieties in what James Rosenau calls a "turbulent world."[11]

Realism, that misleadingly comprehensive catchall category applied to the dominant approaches to international relations scholarship, has been challenged by analyses concerned with emphasizing both the limited capabilities of states in many important spheres of activity and the dangers of defining the relative success of a state as the most important political priority of practitioners of national security. This is a crucial part of the current methodological and philosophical debate within political science and cognate fields about international relations. The limitations of traditional cold war approaches are emphasized by contemporary contributions focusing on culture and identity as well as on critical security studies.[12] This academic debate about identity and what is to be secured has recently, at last, found its way into influential mainstream nonacademic discussions of the future of geopolitics.[13]

The lack of geopolitical consensus, coupled to the broad debate about the concept of security (and it is very important to note their interconnection) leaves room, it seems, for nearly endless discussion of security and how it should be modified, reconceptualized, extended, or addressed in all these ways simultaneously. Many of the critics of national security in the cold war are no longer so obviously marginal to discussions of contemporary dangers. Their ideas often appear in new forms in the debates in what is now rather inadequately called "global civil society." These discussions are not without influence, as the recent international campaign against land mines, among others, has suggested.[14]

Now, perhaps even more so than during the cold war period, the distinction between scholarly analysis and legitimization practice in discussions of security can be seen to be very blurred. Security is usually a political desideratum well before it is an analytical category. In R. B. J. Walker's succinct phrasing, "The forms of political realism that play such a crucial role in the legitimization of contemporary security policies affirm the way things should be far more clearly than they tell us how things are."[15] The arguments about extending or broadening security after the cold war make this point very clear, albeit often accentuating many of the dilemmas of the cold war security discourses.[16] The applicability of much of this academic theorizing to the South, where so much of the contemporary violence actually occurs, further complicates the debate and highlights some of the most important dilemmas.

The argument in this book suggests that all these matters have to be considered together in trying to think about security studies as academic research, policy advice, or pedagogic practice. While academic institutional concerns and debates over concepts of security matter, considering them in the larger political context of contemporary geopolitical changes allows one to better understand both the current state of the field and its place in the academy, and to make suggestions as to how practitioners might rethink what they research and teach as well as the policy advice they suggest to the rich and powerful.

BROADENING THE AMBIT OF SECURITY

From the late 1940s to the late 1980s among the major world powers, national security focused on the military dimensions of the cold war confrontation. But in much of the rest of the world, similar preoccupations dominated state understandings of security. Bradley Klein has convincingly argued that these practices of security were effectively part of the processes of policy coordination whereby American political hegemony was made effective.[17] In the aftermath of the cold war there have been numerous arguments for expanding the remit of security that have challenged the cold war assumptions, but that are also often shaped directly by the engagement with dominant themes from the past.[18]

Preoccupied by matters of military strategy and technical capability, cold war security studies focused on deterrence and the finer points of preventing nuclear war through ensuring that 1939 did not repeat itself. In Moscow those in the security professions worried about a world war caused by the next crisis in international capitalism, thinking that the wars of 1914 and 1939 were unlikely to be the last.[19] Underscored by taken-for-granted geopolitical codes that divided the world into blocs, with danger inherent in the dynamics of the opposing social system, security was a matter of "keeping the bad guys out" by the threat of the use of force. The superimposition of the dynamics of the security dilemma aggravated these geopolitical premises and, given the usually unquestioned assumptions about the political nature of the cold war antagonist, often focused attention on mainly technical matters.

To the critics of cold war politics these conceptions of security seemed unduly constraining, especially in terms of the obvious dangers of military action in a nuclear armed world. States appeared to be endangering the very populations that they were supposedly protecting precisely by trying to ensure national security.[20] These dilemmas were acute when viewed through feminist lenses where women are seen to be especially vulnerable to militarism.[21] But military and direct strategic threats were not the only matters that endangered the states of their citizens. Clearly, to some commentators, a broader definition was needed.

In 1983 Richard Ullman had suggested that

> a more useful (although certainly not conventional) definition might be: a threat to national security is an action or sequence of events that (1) threatens drastically and over a relatively brief span of time to degrade the quality of life for inhabitants of a state, or (2) threatens significantly to narrow the range of policy choices available to the government of a state or to private, nongovernmental entities (persons, groups, corporations) within the state.[22]

When cold war tensions relaxed as the Gorbachev-initiated policies took effect in the Soviet Union in the later years of the 1980s, the critique of national security was extended once more to encompass other matters, especially concerns with the environment. Soviet initiatives at the United Nations and in bilateral negotiations on arms control emphasized a more comprehensive approach to security. Ullman's new definition began finally to be cited in North American discussions, and the ambit of concerns under the rubric of security expanded. While it is important to note that this is not the first time such matters have come under the security rubric—in the late 1940s matters of population and resources were on the agenda; in the 1970s concerns with environment, resources, and, in particular, oil were prominent in American discussions—it is probably fair to argue that they are now more comprehensively incorporated into mainstream discussions in at least North America and Europe, and even in a small way into the declarative security policies of these states.

Numerous items are on the post–cold war security agenda. It is an agenda that is now understood in global terms as a matter for all humanity, although the majority of the authors of such books and articles on security are still probably Americans. As noted in the introduction, paradigmatic here is Davis Bobrow's presidential address to the International Studies Association in 1995. The vast diversity of "new" themes, Bobrow suggested, required rethinking the identity of the security practitioner and reformulating it in terms of medical metaphors to diagnose and prescribe remedies to numerous symptoms.[23] The contents list of the 1998 edition of the textbook *World Security* also suggests the comprehensive nature of these concerns.[24] From chapters on world interests, global dynamics, and great powers, through discussions of conventional arms transfers and the causes of internal conflict, the book goes on to consider violence against women, criminal organizations, socioeconomic disparities, world trading systems, environmental scarcity, and demography, as well as the need for new worldwide institutions. The important argument, made forcefully in its opening chapter by Seyom Brown, is that the traditional understanding of realist politics, that the state was the highest object of concern to the statesperson, is no longer tenable in a world of global interconnections.[25]

Barry Buzan and his colleagues in Copenhagen synthesized many of these themes into a "new framework of analysis" to incorporate the additional dimensions of the concept and some of the critical thinking about security as a political performance.[26] In parallel with other contemporary social theorizing, they emphasize that security is in part a "speech act" that calls into existence a situation of extreme danger requiring extraordinary actions that are understood as such by at least part of the audience to which it is addressed. These authors then go on to link this formulation of security to the broadened agenda of security, suggesting that there are at least five obvious "sectors" in which security is relevant to considerations of world politics. This follows long-standing practice among social scientists to distinguish between society, economy, and politics. The assumptions built into such divisions are not without difficulties, but Buzan, Wæver, and de Wilde suggest that the sectors can be understood in terms of specific types of interaction:

> In this view, the military sector is about relationships of forceful coercion; the political sector is about relationships of authority, governing status and recognition; the economic sector is about relationships of trade, production, and finance; the societal sector is about relationships of collective identity; and the environmental sector is about relationships between human activity and the planetary biosphere.[27]

Given this broadened agenda of matters to be considered in terms of security, the state can no longer be the only factor that is a "referent object" of security. In the case of environmental security, the referent object is the planetary attributes necessary to sustain civilization.

BROADER STILL: HUMAN SECURITY

Other analyses have broadened the concept of security even further.[28] In doing so, they have drawn on senses of the term "security" beyond the specifications conventionally used in international relations. Through the cold war, Western states paralleled their military security concerns with social security programs, understood as providing at least some health and education services that supposedly ensured the basic welfare of all parts of their population. Likewise, income support payments were sometimes considered in terms of the provision of "economic security." The historical dimension of these themes should not be forgotten in all the claims to novel understandings of security.

The impetus for the United Nations was in part to try to ensure that the breakdown of the international economy in the early 1930s that was related to the rise of fascism and Nazism did not reoccur. "There have always been two major components of human security: freedom from fear and freedom from want. This was recognized right from the beginning by the United Nations. But later the concept was tilted in favor of the first component rather than the second."[29] The understanding of widespread political instability as an indirect cause of war also partly underlay attempts to establish the Bretton Woods financial arrangements and a variety of United Nations agencies dedicated to social and economic programs.

The highest profile articulation of these in terms of human security comes from the United Nations *Human Development Report 1994*. The concept of human security has, we are told by the United Nations Development Program (UNDP) authors, at least four essential characteristics. First, it is a universal concern relevant to people everywhere. Second, the components of security are interdependent. Third, human security is easier to ensure through early prevention. Fourth, and perhaps the crucial innovation in this formulation, is the shift of the referent object of security from states to people. This formulation defines human security as "first, safety from such chronic threats as hunger, disease and repression. And second, it means protection from sudden and hurtful disruptions in the patterns of daily life—whether in homes, in jobs or in communities. Such threats can exist at all levels of national income and development."[30]

The argument suggests that the concept of security must change away from cold war and realist preoccupations with territorial security to focus on people's security, and from armaments toward a reformulation in terms of sustainable human development. Demilitarization is obviously part of this agenda, but human welfare broadly conceived is the overall thrust of the concept. As such there are numerous threats to human security, although specific threats are likely to be locale dependent. Nonetheless, in a list that overlaps with the Copenhagen framework, according to the UNDP, threats to human security come under seven general categories: economic, food, health, environmental, personal, community, and political. Food, health, and personal security are additions to the Copenhagen framework, but there is a broad congruity with the other four themes of economic, environmental, societal, and political security.

While many items on the UNDP list of threats to human security are local threats, global threats to human security in the next century are said to include at least six categories, caused more by the actions of millions of people rather than deliberate aggression by specific states. As such they would not be considered security threats under most narrow formulations of security studies. These six are: unchecked population growth, disparities in economic opportunities, excessive international migration, environmental degradation, drug production and trafficking, and international terrorism.[31] These categories and the formulation of human security underlie a growing number of assessments of the current state of world politics and, in particular, recent attempts to encapsulate political agendas for reform of the international system.[32] Being used in such a manner emphasizes the important point that such discussions of security are frequently much longer on political aspiration than they are on analytical clarity.

The multiple extensions of security are usefully summarized by Emma Rothschild in four themes. First ("downwards"), "(T)he concept of security is extended from the security of nations to the security of groups and individuals." Second ("upwards"), "it is extended from the security of nations to the security of the international system, or of a supranational physical environment." Third ("horizontally"), it is extended "from military to political, economic, social, environmental, or 'human' security." Fourth, and in some ways most important for the argument that follows, "the political responsibility for ensuring security (or for invigilating all these 'concepts of security') is itself extended: it is diffused in all directions from national states, including upwards to international institutions, downwards to regional or local government, and sideways to non-governmental organizations, to public opinion and the press, and to the abstract forces of nature and the market."[33] This suggests a multiplicity of extensions that any conceptual analysis will be hard-pressed to accommodate, and that have provoked sometimes intense responses as to the analytical utility or political desirability of such formulations.[34]

RETHINKING SECURITY

In a recent overview of the concept of security, David Baldwin suggests that the lack of a clear definition of the concept of security, which is part of what the whole "rethinking" discussion is about, is because through the cold war period most practitioners were interested in military statecraft and were largely unconcerned with whether what they studied was designated as security, military, strategic, or war studies.[35] Security apparently had utility as a label for an academic practice rather than specifically as an analytical concept. This obviously changed in the 1980s, and after the cold war there has been a growing literature on the subject. More saliently, Baldwin argues that security may no longer be of any use as an analytic concept, having been so widely used for numerous political reasons as to have lost any kernel of meaning that could be elaborated into an academic concept.

Apart from this difficulty of specification, there are two conventional counterarguments to the assumption that extending the security agenda is necessarily an appropriate way to think about contemporary geopolitical issues. First is the concern of traditional realist analysts of national security that the agenda is expanded in a way that dilutes concern with military matters; the primary concern of the military to fight wars may be compromised by competing "threats" that might be better considered by other policy discourses and with policy instruments that are not directly connected to traditional matters of defense.[36] This line of argument suggests that matters traditionally understood as defense and military strategy should remain the appropriate analytical domain of strategic thinking, but that the broadened "security" agenda should be excluded from strategic studies.

The second argument against the broadened agenda for security is ironically the opposite of the first one. It suggests that military approaches are inappropriate tools to tackle many of the new items on the political agenda and that matching appropriate institutions to the new items on the agenda is important. Especially in the case of environmental matters and human rights, traditional military concerns with secrecy and strategic planning are inappropriate for issues where transparency and dialogue, monitoring and cooperation are crucial across international frontiers. The argument here is for a demilitarization or "desecuritization" of many aspects of social life on the assumption that treating them as dangerous invokes exceptional measures rather than dealing with them as routine political and economic matters.[37]

The political ability to specify a threat to a collectivity is obviously an important part of the process of security. The ability to specify danger and mobilize a "we" against a supposedly threatening "them" has long been fundamental to the processes of politics. Carl Schmitt's insight that the distinction between friend and enemy is basic to what politics is about is often clearest in discussions about danger and threats, the subject matter of security.[38] Neorealist scholarly articulations of security sometimes obscure the political dimensions of the matter in detailed technical analyses of weapons systems, defense budgets, or the social propensities of states to warfare. As the discussion of the UNDP security agenda suggests, the debates about rethinking security in the last decade have also often operated to obscure the political dimensions of the matter in constructing normative schemes that turn into political wish lists to secure all manner of things.

One of the key arguments in the postmodern, poststructuralist, and feminist-inspired critiques of conventional international relations thinking in general, and security in particular, has been that the taken-for-granted categories of security are better understood as constitutive of the political.[39] Whatever the impression from the discussions in contemporary literature and the less than helpful designations of these discussions in terms of various "posts," this is not a novel argument, but one that runs back through the debates about modernity and the emergence of sovereign states. Thomas Hobbes and John Locke both wrote about language and understanding as important parts of politics, although these arguments frequently disappear in the oversimplifications of international relations textbooks and the claims to scientific method. The contemporary interests in discourse and theories of representation follow up these themes, making the point in various ways that security is a highly contested political concept precisely because of its location within numerous political difficulties.[40]

The analytical task that arises from the insights of contemporary social theory is not one suggesting the need for yet more conceptual analysis to understand the content and meaning of security better, because security is not just a matter of content. This point is one that Baldwin misses in his dismissal of arguments about security's essentially contested nature. He does so by using a narrow distinction between political and analytical categories that does not allow for an understanding of political language and the social constitution of security as part of the processes that analysis has to engage. Rather, maintaining an understanding of conceptual analysis as only a matter of clarification and stabilization of meaning, he fails to open up his careful and apposite review of Wolfers's earlier arguments about the ambiguity of the concept.[41]

Analyzing its invocation as a political discourse, however, can reveal much about the political "values" that Baldwin argues security is but one among. Understanding security as a "thick signifier," in Jef Huysmans's inelegant phrasing, suggests that it is important to understand the wider order of meaning within which security itself is embedded.[42] Security as a thick signifier "does not refer to an external, objective reality but establishes a security situation by itself. It is the enunciation of the signifier which constitutes an (in)security condition. Thus the signifier has a performative rather than a descriptive force. Rather than describing or picturing a condition, it organizes social relations into security relations." This implies that "security is not just a signifier performing an ordering function. It also has a 'content' in the sense that the ordering it performs in a particular context is a specific kind of ordering. It positions people in their relations to themselves, to nature and to other human beings within a particular discursive, symbolic order."[43] Threats and dangers, and who or what is threatened, are then a matter of politics in particular contexts, rather than of an ontology that can be clarified through conceptual analysis. As Michael Dillon also argues, security is constituted rather than given, even though reality is specified in ways that obscure these processes of representation.[44]

While such theoretical considerations have sometimes been interpreted to suggest that questions of identity and culture need to be brought back into analyses in international relations, Michael Williams argues that identity never left international relations. Rather "a specific conception of identity is in fact constitutive of, rather than missing from, prevailing theories of International Relations and security."[45] A liberal identity has been present and taken for granted in international relations all along, he contends. This is an identity that has roots in the attempts to rethink the politics and knowledge of a very violent period and specifically has to be understood in response to the violence of the Thirty Years' War, the emergence of the Westphalian system, and in the case of British thinking, a response to the violence and disruptions of the civil war in the same period.

Williams suggests that this transformation is both political and epistemological and has the effect of challenging ontological categories profoundly. Thus empiricism is understood as a social practice that, through its skepticism, reduces the provenance of "truth" and in the process allows for individual religious freedom, which has the political consequence of depoliticizing ontological claims to identity that require violent assertion.

> Liberalism sought and represented a transformation of knowledgeable
> practices involving not simply a theoretical innovation of a naive vi-

sion of a natural evolution toward "objective" knowledge, but was part and parcel of an attempt to construct a new set of political institutions and practices within the state, a set of practices which had the question of "security" in the broadest sense at their heart. The new knowledgeable practices of liberalism sought to provide foundations within which political agreement could be obtained and social concord achieved. It sought, above all, to restore a foundation and provide stability to a culture wracked by political conflict and slaughter.[46]

Toleration and the removal of religious identity from citizenship was a political strategy designed to reduce the likelihood of warfare by removing the direct temptations to assert identity violently. It also acted to reduce the efficacy of heroic identities and the glorification of warfare. This was a politics of pacification designed to reduce the dangers of political violence in both public and private realms, freeing the latter to concentrate on wealth accumulation.

While this line of analysis has considerable implications for the methodological discussions within international relations in general, it is important to note that security is understood as individual freedom from political violence and as the precondition for economic activity. Emma Rothschild suggests that the contemporary notion of the security of states came later, particularly from the Napoleonic wars and political upheavals that these caused.[47] This extension of the concept of security in some ways eclipsed the definition of security in individual terms, extending the ambit of security in ways that can simultaneously induce both conceptual confusion and produce policy statements of sweeping generalization that may not offer useful criteria for either clarifying political priorities or choosing the appropriate allocation of state bureaucratic resources. Nor does this discussion present any obvious indication of how scholars might tackle these matters. What should be studied, why, and how is also far from clear in the academic discussions of these things.

Contemporary concerns with human security, and the broadened agenda of political responsibilities for its provision, parallel the earlier liberal politics of the late eighteenth and early nineteenth centuries by refocusing on the safety of the individual. Human security is thus connected to other themes of enlightenment thinking and, in later forms, to human rights. All these liberal themes appear in the recent discussions of global and human security, but liberalism itself is not only a contested tradition, between the more narrowly focused economic doctrines (now often designated neoliberalism) that support the economic supremacy of the rich and powerful, but also a more inclusive tradition that supports an extended understanding of rights, as well as a tradition that has itself long influenced thinking about international relations.[48] Remember also that the cold war international order promoted by the United States was to a substantial degree one premised on "liberal" economic policies but undergirded by military power.[49] The related focus on the individual as the referent object for security is also an important theme in the recent attempts to formulate an explicitly critical security studies.[50]

However, this discussion can be viewed, as much contemporary social theory suggests, in a slightly different way by focusing on the politics of identity and the crucial politicizing question of who we are.[51] David Campbell follows Benedict Anderson in discussing this theme in terms of states as imagined communities.[52] "Central to the process of imagination has been the operation of discourses of danger which, by virtue of telling us what to fear, have been able to fix who 'we' are."[53] While Campbell argues that the United States is the imagined community par excellence, the process works elsewhere also. But especially in the context of the United States, and its multiple formulations of danger, internal and external to the moral order of its political identity, the post–cold war expansion of the ambit of security might better be read as the reemergence of multiple expressions of identity that are the crucial foreign policy making practices now that the dominant cold war code no longer operates as the primary axis of threat. Collective articulations of the dangers of drugs, economic challenges, diseases, rogue states, and an array of environmental dangers are formulated within the same discursive frameworks of danger as were cold war anxieties. Clearly these are also frequently articulated in global terms where the origins of the dangers are attributed to external causes requiring eternal vigilance both at home and abroad.

GEOGRAPHIES OF SECURITY

The greatest enthusiasm for global approaches to security studies and for rethinking the concept comes from the states in North America and Europe whose security situations are understood to be least likely to face direct military threats.[54] Is it politically significant that these discussions are happening where and when they are?[55] If security is, as Bradley Klein has argued more narrowly about strategic studies, tied directly into matters of global political power, what then is the significance of these understandings in particular places?[56]

Many political elites in the poorest states see national security in military terms as their highest priority and the necessary base for constructing a state that might subsequently provide other forms of security. They are supported in this by some modified realist analyses, notably articulated by Mohammed Ayoob, which suggest that states have to be made and nations built first, if necessary by force, prior to attaining the benefits of Western liberal prosperity.[57] This argument is supported by the recent historical construction of states by military force. Many Southern states are more effectively military security organizations than anything else, and insofar as they are, they frequently render their own populations insecure.[58] But:

> [C]ontra Ayoob, one is likely to find the obverse relation between security and the achievement of other societal goals. Said to be mutually constitutive, the practices of civil society and the state in the South Asian case in fact provide a stark rebuttal to the notion that the pursuit of "national" security can create either a more viable, say democratic, polity or an autonomous civil society. Instead, the pursuit of national security undermines both.[59]

Similar arguments can be made in the case of Southern Africa, where the pursuit of national sovereignty provides political elites with powerful arguments for enhancing their wealth and prestige but at the cost of maintaining political and economic systems that are antithetical to the needs of the poorest segments of the population and to the practices of sustainable environmental management.[60]

Large-scale political violence, with a few notable exceptions, in the last few decades has occurred in the South, while many of the weapons come from the North.[61] As the Carnegie Commission on Preventing Deadly Conflict ruefully notes: "Even as governments in many of these states have lost the ability to provide basic services for their populations, they still find ample resources to buy arms."[62] Military organizations are often the dominant organizations where human insecurity is greatest. This is especially true among the poorest states of the South, where the suggestive statistic of the ratio of military expenditure to health spending is often highest, although as Baechler's analysis (discussed in chapter 3) makes clear, there are important exceptions to this generalization, and causation is not necessarily implied in the correlation.[63]

Understanding military forces as the principal security problem for many populations links to larger critiques of militarism and to contemporary analyses of the causes of violent conflict, as well as to the larger agendas of restricting weapons production and utilization as part of the human security agenda. These extensions of security thinking lead away from a focus on interstate warfare and toward an understanding of security in the context of global processes that extend beyond narrow concerns with interstate military and power competition. All these themes are involved in the emergence of the most prominent theme in the broader security agenda discussions in the late 1980s and early 1990s, the theme of the environment. But they also have to be understood in terms of the discourses of insecurity within which they are structured. These are profoundly political issues even when rendered in the most technical appraisals of violence and insecurity.

ENVIRONMENTAL SECURITY

Discussions of the relationships between environment and security didn't start in 1989, although at least in the United States it is fair to say that the topic emerged in its contemporary form then.[64] Against the backdrop of the long summer drought of 1988, alarmist reports of huge tropical forest destruction, especially in Brazil, renewed concern about global climatic change and stratospheric ozone depletion, the relaxation of the cold war, and the drastic rethinking of Soviet security policy, policy discussions in Washington were ripe for some new topics and thinking. Just as Francis Fukuyama was declaring the end of history and the triumph of liberalism, the environment, too, became part of the foreign policy discussion and the focus for discussions of endangerment.[65] In 1989 Norman Meyers published an article linking environment and security in *Foreign Policy*, and Jessica Tuchman Mathews published one in *Foreign Affairs* that suggested that resources and population issues mattered as foreign policy priorities and should be incorporated in a reformulated understanding of security. Mike Renner's Worldwatch paper of that same year also linked environment and security.[66]

In Britain Neville Brown published a paper on climate change and conflict in *Survival*, the journal of the influential International Institute of Strategic Studies; Peter Gleick reversed the process of introducing environment into security considerations by writing about international security in the journal *Climatic Change*. Arthur Westing, the leading researcher on questions of the environmental disruptions caused by warfare, contributed a discussion of a comprehensive formulation of security. Josh Karliner suggested that the environmental difficulties in Central America amounted to a different form of warfare there.[67] National sovereignty and the transboundary responsibilities for the global environment were also the topic for articles.[68] Special issues of the journals *Millennium* and the *Fletcher Forum on International Affairs* followed in 1990. A little later Gwyn Prins introduced the discussion to wider British audiences in a book and television documentary with the memorably apt title of *Top Guns and Toxic Whales*.[69]

Daniel Deudney was quick to pen a paper arguing that all this was not necessarily a good idea. In what has probably become the most cited paper in this whole discussion, he argued that linking security to ecology required a number of serious mismatches of means and ends as well as a misconstrual of the nature and significance of environmentalism.[70] He argued that environmental problems are often diffuse and long-term while wars are concentrated and violent. Polarizing discourses to mobilize populations against identified antagonists is not similar to the kinds of social changes needed to deal with environmental difficulties. With rapid increases in international trade supplying raw materials from a diversity of sources, most resource conflicts were unlikely to lead to warfare. Before the debate had developed very far, one of the arguments Deudney made, that military institutions' frequently dreadful record on environmental matters in the past did not bode well for their handling matters of ecology in the future, was powerfully reinforced by pictures of blazing oil wells in Kuwait in the latter stages of the Gulf War in 1991.

This image appeared on the cover of the issue of the *Bulletin of the Atomic Scientists* that featured a debate between Deudney and Gleick in May that year about linking war and environment.[71] Part of the difference of opinion can be explained by the focus on international organizations and cooperation by Gleick and Deudney's focus on national security institutions, but the disagreement over assumptions that resource difficulties would possibly lead to war was crucial. Lothar Brock and Patricia Mische followed this discussion with chapters in a book on rethinking peace research the following year; Mische advocated extending security discussion to "defend" the environment, and Brock warned of the political dangers of militarizing environmental issues.[72] Matthias Finger argued pointedly that the forms of security that industrialized states pursued were part of the problem that needed to be addressed. Extending security without understanding the roots of modern national security in the destruction of natural environments, he argued, was to badly miss the point, especially when military institutions have often been exempt from environmental legislation on grounds of national security prerogatives.[73]

Other critics extended these arguments in various ways while military representatives suggested that whatever the problems with military actions in the past there was considerable potential for military know-how and technology to make a contribution to studying global change.[74] The question of military damage to the environment has also been a persistent if relatively minor theme in more recent discussions of environmental security. Discussion about how to think about security has continued apace, with advocates of various positions engaging in discussions drawing from a variety of theoretical traditions to rework the basic contentions many times.[75] But as this book shows, in all this considerable discussion neither the geographic premises nor the concept of environment itself has had much analytical attention. These crucial themes and the related matter of the identities secured in the practices and discourses of security, rather than the finer points of the ongoing debate about the concept of environmental security, are the focus in what follows.[76]

From the point of view of scholars and policy analysts interested in either researching the links between environmental phenomena and human insecurity or in providing advice to governments on policy priorities, the most obvious critique of very broad definitions of security is that anything can be judged to be a security threat if humans are harmed by it. Indeed, the human security agenda espoused by the UNDP is more a list of desirable conditions for human beings than it is either an analytical category or a policy program. If all cases where damage to a human being in some form is included, then analysis is impossible. Narrowing the field of inquiry, so the argument goes, is essential if one wishes to either make specific security policy suggestions or use security as an analytical device in empirical research.[77]

The empirical questions that some social scientists have investigated in detail cannot be separated entirely from these conceptual matters, although it is also clear that empirical investigations must take some of the conceptual and ontological categories for granted to allow practical research to be undertaken. Thomas Homer-Dixon, the Toronto-based researcher who led the way in organizing and directing the first major collaborative research project in this field, has often evaded the definitional debate surrounding security to focus on questions of violence and acute conflicts. His first major journal article, published in 1991 in *International Security,* set the terms for much of the empirical research and methodological debates that were to follow.[78] This article also provided a major stimulus to the most high-profile articulation of the environmental security discourse, Robert Kaplan's cover story in the February 1994 *Atlantic Monthly* magazine, examined in detail in the next chapter.

2

The Environment as Geopolitical Threat

Population, when unchecked, increases in a geometrical ratio. Subsistence increases only in arithmetical ratio. A slight acquaintance with numbers will shew the immensity of the first power in comparison of the second.

By that law of our nature which makes food necessary to the life of man, the effects of these two unequal powers must be kept equal.

This implies a strong and constantly operating check on population from the difficulty of subsistence. This difficulty must fall somewhere and must necessarily be severely felt by a large portion of mankind.

THOMAS MALTHUS, *AN ESSAY ON THE PRINCIPLE OF POPULATION*

Every explosion of social forces, instead of being dissipated in a surrounding circuit of unknown space and barbaric chaos, will be sharply re-echoed from the far side of the globe, and weak elements in the political and economic organism of the world will be shattered in consequence.

HALFORD J. MACKINDER, "THE GEOGRAPHICAL PIVOT OF HISTORY"

ONCE AGAIN, THE MALTHUSIAN SPECTER

Robert Kaplan's cover story in the February 1994 *Atlantic Monthly* magazine painted a particularly depressing picture of the future.[1] In "The Coming Anarchy" he argues that much of the world is on a path to violence-ridden anarchy where states collapse and private armies and organized crime establish themselves as effective local administrations. In Mackinder's terms, he clearly suggests that the explosion of demographic and environmental forces has already shattered the weak parts of the political and economic organism.[2] The natural environment is the key villain in the piece. Its degradation has, he asserts forcefully, set off a downward spiral of crime and social disintegration in many places. What is now the case in West Africa will soon spread further as environmental problems generate further migration to urban areas in the underdeveloped world resulting in turn in more social disintegration and ethnic conflict. These issues will become the national security issue for the United States in the next century. The natural environment is thus specified as the threat of the future.

While Kaplan's article generated an angry response from readers who contested his specific accounts of various countries in the letter pages of subsequent issues of the magazine, the themes he wrote about clearly resonated with contemporary American angst about crime, environmental deterioration, and the lack of clear direction to post–cold war security and foreign policy planning. His rhetorically powerful analysis is a high-profile public articulation of contemporary neo-Malthusian themes in post–cold war geopolitical discourse.[3] It parallels much of the rest of the U.S. media coverage in its representations of Africa, and Rwanda in particular, as a place of tribal, hostile, violent Others.[4] It is notable for its pessimism, forceful prose, and the absence of any suggested substantive political remedies to the immanent dystopia.

But Kaplan is not alone. Readers of contemporary international relations literature, foreign policy journals, and magazines of popular political discussion, especially in the United States, have noted that there has been a revival of interest in the themes that concerned Britain's first professional academic economist.[5] Thomas Malthus, the country parson who is widely memorialized for his pessimism about humanity's lot, a fate due largely to our supposed predilection for breeding faster than we can improve our capabilities to feed ourselves, is again in vogue in post–cold war policy discussions, but now his theories are often linked to themes of environmental degradation and to some of the traditional themes of geopolitics.

In the mid-1990s, as I wrote the first draft of what subsequently became the introduction to this chapter, my radio was broadcasting a documentary on the politics of Algerian migration to France with alarmist projections of sixty million unemployed Arab youths in North Africa providing a security threat to Europe. On the local television station, that same week's episode of the *X-Files* was set in the context of U.S. military incarceration of Haitian refugees. The question posed by these and similar contemporary popular geopolitical narratives is often, to use Malthus's phrasing, a matter of precisely where the "difficulty" will fall.

Against the backdrop of the major UNCED meetings in Rio de Janeiro in June 1992, none of this renewed concern with population as a political factor is perhaps very surprising. But when this theme is linked, as Kaplan explicitly does, to the more general concerns about environment as a security threat, these arguments become important in the political processes of foreign and security policy formulation in states in the North. Foreign and security policy prescriptions depend in part on how the questions of appropriate policies are practically understood within the larger geopolitical discourses and their interpretations of contemporary geopolitical order. The same is true about themes of the environment in international political discussions and policy formulation.[6]

The more popular media discourses in discussions of the future of environmental factors in security policy are not nearly so sophisticated, but they are likely to get political attention when published as a cover story in a prestigious upmarket magazine like the *Atlantic*. Kaplan was taken seriously in the White House, given his track record as a travel writer and war correspondent with a knack for getting into conflict areas. In particular, his book *Balkan Ghosts* reportedly had considerable influence on President Clinton's policies in Bosnia.[7] Journalists reporting on the theme make clear that Kaplan's "Anarchy" article has also been influential in Washington foreign policy–making circles.[8] It was specifically cited by President Clinton in a speech soon after its appearance and "became a practically *de riguer* citation among Cabinet members appearing before Congress."[9]

While Kaplan's article did not initiate the policy process considering the links between security and environment, it apparently "played a catalytic role in bringing the environment-conflict thesis to the attention of the highest levels of the Clinton Administration and the larger Washington policy community."[10] Subsequently, there has been continued interest in the topic in Washington, where the Woodrow Wilson Center has taken the lead in promoting and publishing policy and scholarly thinking about environment, conflict, security, and population. Closely related matters of state failure and violence, and the priorities for American foreign policy in dealing with environmental collapse, have generated numerous books and reports that, in part, are the subject for later chapters.

MALTHUS AND MACKINDER

In many ways none of this is very new. In England, in the years following Malthus's initial publication during the transformations of the industrial revolution and in the aftermath of the American and French revolutions, there were widespread concerns among the political elites and in the emergent middle classes about political order and the fear of the mob as a destabilizing social factor. As Michel Foucault has argued, it was in the period immediately prior to Malthus that the conception of population as an object to be controlled, manipulated, and managed by states clearly emerged as an important factor in modern modes of governmentality.[11] (As noted in chapter 1, this was also the period when security was extended from individuals to states.) The extension of these modes of governmentality to the global population and its environment, which provides the means of subsistence, was a recurrent theme of the twentieth century, at least in the discussion of the policy elites who concerned themselves with "global" problems and questions of world political order. Historical research into the colonial management of resources in the nineteenth century shows that contemporary discussions are in many ways a continuation of these earlier practices.[12]

Malthusian arguments have also long been used as a political strategy to avoid directly discussing the structural causes of inequality.[13] Most important in their efficacy is the simple but very powerful assumption in modern economic discourses that general scarcity is the ontological condition of humanity.[14] Coupled to the growing awareness of life on a small planet, and hence the inevitability of certain limits on humanity's actions, this assumption frequently obscures detailed discussions of the particular circumstances of poverty. It does so by naturalizing scarcity within a larger managerial mode of reasoning that obscures the political economy of both resource-use decision making and the artificial construction of commodity scarcity in many market systems. A general assumption of scarcity often works as an ideological move to preclude the necessity of probing into distributional questions.

Fear of overpopulation and social hardship was a recurring political theme through the cold war, albeit one that was much less prominent than concerns with superpower rivalry. Harrison Brown's book on *The Challenge of Man's Future* in the early 1950s was a discussion of then contemporary Malthusian themes.[15] A generation later Paul Ehrlich published *The Population Bomb*, which generated considerable controversy with its dire predictions of future catastrophe.[16] Similar attention was paid to *The Limits to Growth* study in the early 1970s.[17] In a paper written in response to this period of neo-Malthusianism, David Harvey points ironically to a 1948 text that argued the inevitability of massive famine in China and the impossibility of geometric population growth there in the 1950s, just before the Chinese population began its dramatic increase.[18] The only famine that did transpire, in the late 1950s, came as a result of economic and political dislocations, not any "natural" limitations.

Following the much publicized African famines of the 1980s, Paul Ehrlich returned to his earlier themes of population growth in a new book called *The Population Explosion*, where he argued that the "bomb" he warned of earlier had now exploded, with huge numbers of people dying each year from hunger and hunger-related diseases.[19] *Beyond the Limits* was published in 1992 as a sequel to *The Limits to Growth*, suggesting policy options to be taken to prevent "overshoot" and collapse by working toward a sustainable society.[20] While estimates of how many people the planet can feed vary widely dependent on assumptions about technology, diet, distribution of wealth, water resources, and calculations of the availability of arable land, the logic of this type of thinking suggests that disaster will occur as natural limits are reached.[21] Many of these themes have also appeared fairly regularly in large-circulation American magazines since the beginning of the cold war. Concerns about population growth and war as a result of expansionist policies often reflected Second World War themes. The environmental limitations on population increase were an important theme in the early 1970s following the first Earth Day and the limits to growth debate. African population growth, however, only became a regular theme in popular discussions of population in the 1980s, often within a "humanitarian" focus on famines.[22]

Given these themes, Kaplan is in some ways continuing long-established lines of argument. But his powerful articulation of environment as the cause of threats to national security has updated Malthusian themes and brought the environmental security policy discussions forcefully to the attention of a wider public than had previously been the case. In doing so Kaplan revisits many of the geopolitical assumptions in security thinking and in specifying the environment as a threat. This use of specific geopolitical assumptions to frame the demographic and related environmental dimensions in post–cold war security thinking is a focus in what follows in this chapter and the rest of the book.

In the case of neo-Malthusianism and the more general policy discourse of environmental security, the threat is often at least partly from somehow external natural or environmental phenomena. More specifically, Kaplan's essay can be read as an analysis of, in Gearóid Ó Tuathail and T. W. Luke's terms, the "wild" zones of the new geopolitical (dis)order where the potential for disruptive incursions into the "tame" zones of postmodern prosperity requires their containment, if necessary by military force.[23] As such the political specifications of particular places in the neo-Malthusian literature follow some of the earlier patterns of environmental determinism that were important in cold war geopolitical thinking, but with the environment now more directly understood as an immediate threat rather than as an indirect historical cause.[24]

But a careful reading of Kaplan's article makes clear that the geopolitical formulations in American political discourse are not simply a continuation of cold war themes. The new danger of environmental degradation is accentuated here, as are demographic concerns, while old concerns about access to strategic resources are often downplayed or ignored. Africa, in particular, is now understood not as a security commodity, which is significant as a place of superpower rivalry and mineral supplies, but as a source of political instability that may, if unchecked by security measures, spread further afield to threaten areas of Northern affluence.

What is also very obvious in Kaplan's text, as has so often been the case in geopolitical reasoning, is the use of geopolitical shorthand in the form of simplified geographs to reduce complex geographical realities to reified entities or spatial metaphors of supposedly stable geographical places that can then become the objects in the narratives used to formulate foreign policy. In this way, "the policy of making foreign," of designating external others as different and potentially threatening, has carried its key practices over from the cold war period.[25] But now, in an ironic reprise of earlier American cultural themes of a hostile nature that needed to be tamed, domesticated, and rendered benign by colonization of the frontier, the environment has been specified as that which is foreign and threatening.[26] Metaphors of wars with nature are not new, but the explicit linkage of military metaphors of nature as a hostile force with geopolitical threats to national security gives these themes a new and potentially ominous twist. In Kaplan's terms:

> It is time to understand "the environment" for what it is: *the* national-security issue of the early twenty-first century. The political and strategic impact of surging populations, spreading disease, deforestation and soil erosion, water depletion, air pollution, and possibly, rising sea levels in critical overcrowded regions like the Nile Delta and Bangladesh—developments that will prompt mass migrations and, in turn, incite group conflicts—will be the core foreign-policy challenge from which most others will ultimately emanate, arousing the public and uniting assorted interests left over from the Cold War.[27]

KAPLAN'S "COMING ANARCHY"

Kaplan's article pulls no punches in its pessimistic vision of environmentally induced social collapse, spreading disease and crime. With armed gangs of "technicals," inspired by "juju spirits" in West Africa, and the widespread collapse of social order in Asia and Yugoslavia, the nation-state is, he argues, quickly becoming a political formation of the past, and sovereignty is now a dated fiction derived from the cartographic practices of another era.

The magazine's designers powerfully reinforce the message. The front cover illustration shows a crumpled map of the world starting to burn on a wood floor, the flames rising into words superimposed on the wall behind. In bold capitals they ominously announce:

> The coming anarchy: Nations break up under the tidal flow of refugees from environmental and social disaster. As borders crumble, another type of boundary is erected—a wall of disease. Wars are fought over scarce resources, especially water, and war itself becomes continuous with crime, as armed bands of stateless marauders clash with the private security forces of the elites. A preview of the first decades of the twenty-first century.

The article is accompanied by stark photographs. The opening pages depict armed soldiers walking past human skeletal remains in Liberia. Photographs of roadside warnings of "killing zones" in Sierra

Leone, of mass graves in Bosnia, and Kurdish guerrillas in Turkey are followed by pictures of human corpses, the consequences of violent retribution in Liberia and Vukovar. Pictures of "the press of population," showing buses amid crowds in Lagos and people doing their washing in an Abidjan lagoon, as well as other photographs of Southern cities, suggest overcrowding. The final photograph is of looters in the post–Rodney King police officer trial riots in Los Angeles, suggesting that the scenes in the earlier depictions were intimations of things to come in the United States. The theme of ethnic conflict is prominent.

Kaplan starts with West Africa where he argues that crime is the order of the day, or rather the order of the night, when what tentative authority governments have dissipates as youthful criminals take to the streets. We are told that organized crime is related to the collapse of the nation-state and the rise of demographic and environmental stresses. Drug cartels and private security forces take over where social stress has led to the collapse of more conventional political order. To Kaplan this is clearly the future of global politics, a specter that confronts "our" civilization and one that conjures up "Thomas Malthus, the philosopher of demographic doomsday, who is now the prophet of West Africa's future. And West Africa's future, eventually, will also be that of most of the rest of the world."[28]

Kaplan suggests that Nigeria may try to assume hegemonic power in the region as French influence and interest declines. On the other hand, Nigeria may fall apart in a cataclysm "that could make the Ethiopian and Somalian famines pale in comparison."[29] He fails to note that Nigeria had already been through such a cataclysm a generation earlier when Biafra seceded in a bloody civil war notable for the mass starvation of the Biafran population. Picking up on another theme in the contemporary popular geopolitical imagination, the spread of deadly diseases, Kaplan portrays them, and new forms of antidote-resistant malaria in particular, as an emerging impenetrable barrier closing the whole African continent off from the rest of the world even as its internal state boundaries crumble.[30] The only exceptions to this exclusion by the "wall of disease" are likely to be coastal trading posts.

This introduces the environmental theme framed in terms of extensive shantytowns on the urbanizing coast of West Africa. "In twenty-eight years Guinea's population will double if growth goes on at current rates. Hardwood logging continues at madcap speed, and people flee the Guinean countryside for Conakry. It seemed to me there that here, as elsewhere in Africa and the Third World, man is challenging nature far beyond its limits, and nature is now beginning to take its revenge." But quite what the mechanism is that drives the migration is not explained; the text merely suggests that it is related to deforestation. Africa may, he suggests, be like the Balkans one hundred years ago, a harbinger of an old (imperial) order collapsing and giving way to nations based on tribe. But a century later the analogy contains a fundamental difference: "Now the threat is more elemental: *nature unchecked.*"[31]

Environmental scarcity is the first of the concepts that one must look at to understand Kaplan's new world. It is linked to cultural and racial clashes, geographic "destiny" and the transformation of warfare. Looking in turn at these themes allows Kaplan to sketch out the map of the new political situation. Of prime importance to all these matters is the environment. In the pivotal passage in his article, reproduced at the end of the last section, he draws on themes from the more pessimistic environmental security literature to argue that the environment is the national security issue of the near future.[32] This is no small claim. It suggests that the fate of modern states is now tied directly to the fate of environments around the world. Ecological disruptions are now to be feared, the environment understood as "a hostile power," as the section heading in his article immediately prior to this quotation puts it.

Kaplan suggests that Thomas Homer-Dixon's 1991 *International Security* article, "On the Threshold: Environmental Changes as Causes of Acute Conflict," may turn out to be the framework for future policy, similar to George Kennan's famous paper in *Foreign Affairs* in 1947, which became the rationale for the cold war containment policy.[33] The thrust of Homer-Dixon's article leads Kaplan to think that growing scarcity of resources, coupled with increasing population numbers, may lead to social pressures, increased migration, environmental refugees, and intergroup conflict in many places. According to Kaplan, Homer-Dixon's research can be interpreted to suggest that the environmental degradation in the developing world "will present people with a choice that is increasingly among totalitarianism (as in Iraq), fascist-tending mini-states (as in Serb-held Bosnia), and road warrior cultures (as in Somalia).[34] The implication is that all these developments threaten political stability and hence, at least indirectly, the security of Northern states. Environmental degradation may well lead to war.[35]

Kaplan also attributes to Homer-Dixon the bifurcation of the world into two areas, one of affluent Northern suburbia, where Hegel and Fukuyama's "Last Man" will be able to master the coming crisis, the other of impoverished Southern environmental degradation, where Hobbes's "First Man" is "condemned to a life that is 'poor, nasty, brutish, and short.'"[36] Kaplan uses the metaphor of the affluent in a stretched air-conditioned limousine driving through the potholed streets of New York City while the rest of the world's population outside the automobile is going the other way, presumably on foot. The world is being divided into two camps: the vast majority, rapidly increasing numbers of the poor, on one side, and the small minority of the rich, who can distance themselves from environmental destruction, on the other.

The clashes between groups that are likely to result from environmental degradation–induced identity conflicts are, Kaplan argues, most probably going to occur along lines of tribal and cultural fracture. In making this case he uses Samuel Huntington's much cited *Foreign Affairs* article, "The Clash of Civilizations," which suggested that long-term cultural divisions were likely to determine the pattern of post–cold war geopolitics. Kaplan argues that because Huntington's argument is painted with such a broad brush some of the details are inaccurate.[37] The clashes in the Caucasus are a matter of cultural identity and Turkish versus Iranian civilizations, rather than a clear battle between the forces of Christianity and Islam as Huntington's thesis suggests. Kaplan points to the continued struggles between the Turkish state and the Kurdish population in eastern Turkey as a contest of great importance for the future of the Middle East, not least because of the presence in this region of major Turkish hydroelectric projects that control crucial water flows into Syria and Iraq.

These specifications of identity in terms of cultures link the text to another theme of classical geopolitics, the focus on "organic communities" as the preferred political communities. As Ó Tuathail notes, Mackinder's political thinking, often remembered in the terms of the introductory epigraph to this chapter as relating to matters of geopolitics, a term Mackinder didn't like, is perhaps better understood in terms of conservative nostalgias for persistent cultural identities that support political stability.[38] The organic assumption of unchanging cultural identities plays into support for clan, tribe, and nation and becomes particularly powerful when coupled to claims to territory and sovereignty. As with Huntington's analysis, eternal social essences and identities are invoked in the face of dramatic social and political change. For Kaplan, only Huntington's scale is wrong; politics is about geopolitical identities that suggest permanent fissures between potentially warring parties.

But an autonomous Kurdish entity is not the solution for Turkey, where Kurds are spread throughout the state. Kaplan suggests that identity and territory are not so closely related as the mapmakers of states suggest in their publications. But the implications of this for the possible future of Turkey is left unclear, hinting at widespread criminal or terrorist violence. All of which brings Kaplan to his reconsideration of the likely future of warfare. Here his guide is Martin van Creveld's strategic ruminations about the changing forms of warfare.[39] Kaplan paints a picture of the return to medieval-type loyalties and warfare between local warlords, families, and tribes. These involve the services of mercenaries that operate with a minimum of concern for local populations or their resources. In Kaplan's reading of van Creveld, "technicals" and militias, terrorists and mercenaries, rather than state-organized, centrally commanded regular armies, are the likely form of future combat in the degraded environments of underdeveloped regions. Crime and war will merge. Sophisticated technologies will be used for very unsophisticated purposes as rivalries turn violent and state armies dwindle and disappear as the social infrastructure dissolves. How do sophisticated technologies, and weapons in particular, produced far from these battlefields get into the hands of these populations? Kaplan never considers the question.

Kaplan ends his article by arguing that coherent national states are a fading political phenomenon, which conventional political cartographies no longer accurately represent, and speculating on the future of India and Pakistan as their burgeoning populations, with long histories of collective violence, face the future on a dwindling resource base. Add to this speculations about global climate change and the future of political order in states like Egypt, and the potential for drastic political upheaval seems huge. Even the United States may not survive given its ethnic tensions and individualist culture.

These tensions might well be aggravated by African disasters as African Americans demand American actions to provide help to stricken populations. The final few paragraphs comment on the author's return to the United States after his research trip for this article, where he saw laptop computer–equipped businesspeople at Kennedy Airport on their way to Tokyo and Seoul. No such people were boarding planes to Africa. The suggestion is once again of two worlds with little connection.

Some months after the article's publication, political violence tore Rwanda apart, and media reports of "tribal" slaughter apparently confirmed Kaplan's nightmarish vision.[40] The stark prose and violent images in Kaplan's article capture the alarmist themes of contemporary neo-Malthusianism. While other articles in policy journals and books by authors as prominent as Paul Kennedy discuss these demographic and environmental themes, Kaplan's article is significant in the bluntness with which he gives these themes widespread popular exposure.

KAPLAN'S GEOPOLITICAL IMAGINATION

However, the world is not quite so conveniently simple as Kaplan's popularization of environmental degradation as the key national security issue for the future suggests. His article, for all its dramatic prose and empirical observation, is vulnerable to numerous critiques. Read as a cultural production of considerable political importance, it is fairly easy to see how the logic of the analysis, premised on eyewitness empirical observation and drawing on an eclectic mixture of intellectual sources, leaves much of significance unsaid. But the impression, as has traditionally been the case in geopolitical writing, generated from the juxtaposition of expert sources and empirical observation is that this is an objective, detached geopolitical treatise.[41] The focus in what follows is on the political implications of the widely shared geopolitical assumptions that structure this text and ultimately render the environment as a threat.

The most important geopolitical premise in the argument posits a "bifurcated world," one in which the rich in the prosperous "posthistorical" cities and suburbs have mastered nature through the use of technology, while the rest of the population is stuck in poverty and ethnic strife in the shantytowns of the underdeveloped world.[42] The closing image in his text, of the affluent New York airport with businesspeople flying to Asia but not to Africa, is very strongly reinforced throughout the article by the sharp contrast between the advertisements for luxuries and the violent imagery of the photographs and the themes in the text.

Insofar as politics is defined in terms of the articulation of discourses of danger, Kaplan's analysis can be read in terms of a persistent textual dualism between postmodern consumer aspirations and fear of "reprimitivized" violence and environmental degradation. The presentation of a bifurcated world is powerfully reinforced by the dramatic juxtaposition of the advertisements with the images and content of the text. All the advertisements suggest the symbols of consumer affluence: three are for automobiles, one for gin, two for stereophonic audio equipment, one for a book club, and another for compact discs. Nothing unusual here. But on closer inspection these advertisements speak volumes about the geopolitics of the contemporary world. Where the article uses the metaphor of stretch limousines for the affluent, driving over potholed streets in New York, the automobile advertisements show the luxury interior of one vehicle and another parked beside a traditional brick house in a state of bucolic bliss. The Saab automobile advertisement, stretching over three pages, emphasizes the achievements of high-technology engineering.

But the juxtaposition of the two worlds of aspiration and fear can be taken further. The article talks of non-Western cultures in conflict and of slums that are described as so appalling that not even Charles Dickens would give them credence, while the book-of-the-month club advertisement is for a twenty-one-volume collection of Dickens's works. The advertisement for a Bose radio is focused on a Stradivarius violin. The advertisement for a Sony CD player shows a grand piano and a Sony scholarship–winning Juilliard School pupil, cultural artifacts far removed from juju spirits, animism, or even Islam. The appreciative student pianist endorsing Sony contrasts dramatically with the mention in the text of Solomon Anthony Joseph Musa, a coup leader in Sierra Leone who, it is claimed, "shot the people who had paid for his schooling, 'in order to erase the humiliation and mitigate the power his middle class sponsors held over him.'"[43] The final advertisement, for Columbia House compact discs, focuses, in a truly bizarre irony, on the history of the blues.

Perhaps most geopolitically revealing, however, is the advertisement for "Bombay Sapphire Distilled London Dry Gin." The juxtaposition of Bombay and London, along with the image of Queen Victoria on the label on the bottle, suggests the legacy of colonialism and the commercial advantages gained by European powers in earlier geopolitical arrangements and, in particular, the access to raw materials in the colonies for use in the production of all manner of commodities for consumption in the metropoles of empire. In all of Kaplan's article, such matters of international trade are barely mentioned. The "wall of disease" may bar many foreigners from all except some coastal "trading posts" of Africa in the future, but the significance of what is being traded, and with what implications for the local environment, is not investigated. "Hot cash," presumably laundered drug moneys from these states, apparently does flow to Europe, we are told, but this has significance only because of the criminal dimension of the activity, not as part of a larger pattern of political economy. While the lack of businesspeople flying to Africa is noted, comments about the high rate of logging are never connected to the export markets for such goods or to the economic circumstances of indebted African states in the early 1990s that distorted local economies to pay international loans and meet the requirements for structural adjustment programs.[44] Logging continues apace, but it is apparently driven only by some indigenous local desire to strip the environment of trees, not by any exogenous cause. A focus on the larger political economy driving forest destruction would lead the analysis in a very different direction, a direction not taken by the focus on West Africa as a quasi-autonomous geopolitical entity driven only by internal developments.

Political violence and environmental degradation are not related to larger economic processes anywhere in this text. This is not to suggest that the legacy of colonialism, or the subsequent neocolonial economic arrangements, is solely to blame for current crises, although the history should not be ignored, as Kaplan does. It is to argue that these sections of Kaplan's text show a very limited geopolitical imagination, one that focuses solely on local phenomena in a determinist fashion that ignores the larger transboundary flows and the related social and economic causes of resource depletion. Kaplan ignores the legacy of the international food economy that has long played a large role in shaping the agricultural infrastructures and the nutritional levels of many populations of different parts of the world in specific ways.[45] He also ignores the impact of the economic crisis of the 1980s and the often deleterious impact of the debt crisis and structural adjustment policies. He completely misses their important effects on social patterns and rural women, who are most vulnerable.[46]

Ironically, given his repeated comments about the inadequacies of cartographic designations of state boundaries in revealing cross-border ethnic and criminal flows, he never investigates their similar inadequacies for understanding economic interconnections that ought to be crucial to his specification of various regions in Malthusian terms.[47] This omission allows him to attribute the failure of societies to purely internal factors. Once again the local environment is constructed as the cause of disaster without any reference to the historical patterns of development that may be partly responsible for the social processes of degradation.[48]

Given the focus of most Malthusians on the shortage of "subsistence" and resources in general, there is remarkably little investigation of how the burgeoning populations of various parts of the world are actually provided for, either in terms of food production or other daily necessities. Despite accounts of trips across Africa by "bush-taxi," agricultural production remains invisible to Kaplan's eye-witness. While cities are dismissed as "dysfunctional," the very fact that they continue to grow despite all their difficulties suggests that they do function in many ways. Informal arrangements and various patterns of civil society are ignored. People move to the cities, but why is never discussed in this article; imprecise references to degraded environments and the world soil degradation map on Homer-Dixon's office wall are all that is offered. There is no analysis here of traditional patterns of subsistence production or how they and access to land may be changing in the rural areas, particularly under the continuing influence of modernization.[49]

While it is made clear that traditional rural social patterns fray when people move to the very different circumstances of the city, the reasons for migration are never investigated. In Homer-Dixon's language, absolute scarcity is assumed and the possibilities of relative scarcity, with the negative consequences for poor populations due to unequal distribution or the marginalization of subsistence farmers as a result of expanded commercial farming, are never considered.[50] Why Malthus, in particular, should be the prophet of West Africa is far from clear. There is disease and crowding in the shantytowns of many cities, a phenomenon that is not new, but not all the new urban populations are dispossessed forest dwellers or refugees from criminal activities.

The focus on environment as the key factor in triggering violent changes is not entirely consistent with Kaplan's arguments elsewhere about the cohesive force of Islam, identified ironically in a few places, given the usual orientalizations in practice when discussing Islam, as a Western religion.[51] His discussion of Turkey suggests that while urbanization is occurring rapidly, social cohesion and resistance to crime is being maintained by Islam, even as new geopolitical identities are being forged in the slums. While he suggests that these identities may transcend the force of Islam in the ongoing conflict between Turks and Kurds, his emphasis on nonenvironmental factors of social cohesion suggests that his argument is perhaps more concerned with traditional matters of ethnic identity and "civilizational clashes" than with environmental degradation.

Here resurgent cultural fears of the Other and assumptions about the persistence of cultural patterns of animosity and social cleavage are substituted for analysis of resources and rural political ecology. Precisely where the crucial connections between environmental change, migration, and conflict should be investigated, the analysis turns away to look at ethnic rivalries and the collapse of social order. The connections are implied, not demonstrated; the opportunity for detailed analysis is missed; and the powerful rhetoric of the argument retraces familiar political territory instead of looking in detail at the environment as a factor in social change. In this failure to document the crucial causal connections in his case, Kaplan ironically follows Malthus, who relied on his unproven key assumption that subsistence increases only at an arithmetic rate in contrast to geometric population growth.

Political angst about the collapse of order is substituted for an investigation of the specific reasons for rapid urbanization, a process that by default is rendered as a natural product of demographic pressures. This unstated naturalization then operates to support the Malthusian fear of poverty-stricken mobs, or in Kaplan's terms, young homeless and rootless men forming criminal gangs, as a threat to political order. Economics becomes nature, nature in the form of political chaos becomes a threat, and the provision of security from such threats thus becomes a policy priority. In this way "nature unchecked" can be read directly as a security threat to the political order of postmodernity.

GEOPOLITICS, MALTHUS, AND KAPLAN

Kaplan explicitly links the Malthusian theme in his discussion of Africa to matters of national security; a clear, external, threatening dimension of crime and terrorism necessitates policy practices of security and strategic thinking. The logic of a simple Malthusian formulation is complicated by the geographical assumptions built into Kaplan's argument, while he has simultaneously avoided any explicit attempt to deal with the political economy of rural subsistence or contemporary population growth. Thus, in his formulation, the debate is shifted from matters of humanitarian concern, starvation, famine relief, and aid projects and refocused as matters of military threat and concern for political order within Northern states.

What ultimately seems to matter is whether political disorder and crime will spill over into the affluent North. The affluent world of the *Atlantic* advertisements, with their high-technology consumer items, is implicitly threatened by the spreading of "anarchy"; already American inner cities are plagued with violent crime. Kaplan suggests a specific geopolitical framework for security thinking: the United States may become more fragmented and Canada may dissolve following the secession of "Quebec shorn of its Northern resource hinterland." He even argues that Quebec, as a supposedly culturally homogenous society, may end up being the most stable region of North America.

What cannot be found in this article is any suggestion that the affluence of those in the limousine might in some way be part of the same political economy that produces the conditions of those outside. This connection is simply not present because of the spatial distinctions Kaplan makes between here and there. He notes the dangers of the criminals from there compromising the safety of here, but never countenances the possibility that the economic affluence of here is related to the poverty of there. The spatial construction of his discourse precludes such consideration, but some of his factors violate the integrity of cartographic boundaries.

Although Kaplan is particularly short on policy prescription in his *Atlantic* article, his reworked Malthusianism does have some clear political implications. Instead of repression and the use of political methods to maintain inequalities in the face of demands for reform, Kaplan's implicit geopolitics suggests abandoning Africa to its fate. If more Northern states withdraw diplomatic and aid connections and, as he notes, stop direct flights to airports such as Lagos, it may be possible to isolate this troubled region. If contact is restricted to coastal trading posts, then the "wall of disease" will become a wall of separation keeping non-Africans out and restricting the migration of Africans. Once again security is understood in the geopolitical terms of containment and exclusion.

In a subsequent article in the *Washington Post,* Kaplan explicitly argues against U.S. military interventions in Africa.[52] He suggests that intervention in Bosnia would do some good because the developed nature of the societies in conflict there allows some optimism that a political settlement is workable. The chances of intervention having much effect in Africa are dismissed because of the illiterate, poverty-stricken populations there. However, the pessimism of the *Atlantic* article is muted here by a contradictory suggestion that all available foreign policy money for Africa be devoted to population control, resource management, and women's literacy. These programs will, Kaplan hopes, in the very long term resolve some of the worst problems, allowing development to occur and "democracy" to emerge eventually.

The ethnocentrism of the suggestion that Africa's problems are solvable in terms of modernization is coupled to the implication that West Africa is of no great importance to the larger global scheme of power and economy and therefore can be ignored, at least as long as the cultural affinities between Africans and African Americans don't cause political spillovers into the United States. In this geopolitical argument Kaplan parallels Saul Cohen's geopolitical designation of sub-Saharan Africa as part of a "quartersphere of marginality," consigning it to irrelevance in the post–cold war order.[53] Precisely this marginalization has long been of concern to many African leaders and academics. But in stark contrast to Kaplan, in the early 1990s many Africans emphasized the need to stop the export of wealth from their continent and to draw on indigenous traditions to rebuild shattered societies and economies.[54]

Spatial strategies of containment are a long-standing component of security thinking. Given the specification that the political turmoil is caused internally, cutting anarchy-ridden regions loose in the hopes that their chaos will remain internal makes sense in an argument that constructs these places as clearly external to the political arrangements that one wishes to render secure from threats. Also given the startling failure in this analysis to consider that matters of international economics are a possible cause for some of the phenomena that are involved in the dissolution of political order, no sense of external responsibility applies.

Kaplan deals with deterritorialized phenomena when they suit his argument, but conveniently ignores transboundary flows when they do not fit his cartographic scheme. They suit it here because they emphasize political violence and threats across frontiers that are in some cases disappearing. Large-scale geopolitical isolation as a cordon sanitaire might work as a Western security strategy in these circumstances; it seems less likely to help Africans, but that is not high on Kaplan's scheme of priorities. But to advocate these solutions is once again to specify complex political phenomena in territorial terms, a strategy that is, as John Agnew argues, falling into the familiar "territorial trap" in international relations thinking where boundaries are confused with barriers, and flows and linkages are obscured by the widespread assumption that autonomous states are the only actors of real importance in global politics.[55]

There is an ironic twist in Kaplan's geopolitical specifications of "wild zones." He argues that they are threats to political stability and, in the case of Africa, probably worth cutting loose from conventional political involvement. In his subsequent *Washington Post* article, he argues against military interventions in Africa on the basis of their uselessness in the political situation of gangs, crime, and the absence of centralized political authority. His suggestions imply that interventions are only considered in terms of political attempts to resolve conflicts and provide humanitarian aid. In this assumption Kaplan is at odds with cold war geopolitical thinking. While ignoring the political economy of underdevelopment as a factor in the African situation, he also ignores the traditional justifications for U.S. political and military involvement in Africa and much of the Third World. Through the cold war and immediately afterward, these focused on questions of ensuring Western access to strategic minerals in the continent.[56] But Kaplan ignores both these economic interconnections and their strategic implications, preferring an oversimplified geopolitical specification of Malthusian-induced social collapse as the focus of concern.

The specification of danger as an external natural phenomenon works in an analogous way to the traditional political use of neo-Malthusian logic. Once again threats are outside human regulation, inevitable and natural in some senses—if not anarchic in the neo-realist sense of state system structure, then natural in a more fundamental sense of nature unchecked. By the specific spatial assumptions built into his reasoning, Kaplan accomplishes geopolitically what

Malthusian thinking did earlier in economic terms. Coupled to prevalent American political concerns with security as internal vulnerability to violent crime and external fears of various foreign military, terrorist, economic, racial, and immigration threats, Kaplan rearticulates his modified Malthusianism in the powerful discursive currency of geopolitics. His themes fit neatly with media coverage of Rwanda and Somalia, where his diagnosis of the future seemed to be already occurring. Understood as problems of tribal warfare, such formulations reproduce the earlier tropes of primitive savagery.

As other commentators on contemporary conflict have noted, detailed historical analysis suggests that the formation of "tribes," and many of the tribal wars that European colonists deplored, were often caused by the sociological disruptions triggered by earlier European intrusions.[57] Denial or failure to understand the causal interconnections of this process allowed for the attribution of savagery to Others, inaccurately specified as geographically separate. Kaplan notes that the disintegration of order is not a matter of a primitive situation but, following van Creveld, a matter of reprimitivized circumstances in which high-technology tools are used for gang and tribal rivalries. But the economic connections that allow such tools to become available are not mentioned. Thus reprimitivization is specified as the indirect result of environmental degradation, a process that is asserted frequently but not argued, demonstrated, or investigated in any detail. Once again, geopolitical shorthand is substituted for detailed geographical analysis. The irony of the policy discourse of geopolitics, as the antithesis to detailed geographical understanding, is in play once again in this text, although this time with environment as a reified concept.

3

Environment, Conflict, and Violence

It is the same cultural factors and behavioral patterns that lead to environmental degradation on the one hand and armed conflict on the other: competing interests in using renewable resources, scarcity and pollution through excessive exploitation, unclear or competing legal structures and rights of tenure, and the political mobilization of collective actors who are bound up in struggles over distribution or their own defense. They are the expression of deep social changes affecting the rural population of developing countries.
GUNTHER BAECHLER, "ENVIRONMENTAL DEGRADATION
IN THE SOUTH"

RESEARCHING KAPLAN'S ASSUMPTIONS

The key source for Robert Kaplan's assumptions about environmental degradation leading to conflict is Thomas Homer-Dixon's 1991 paper in *International Security*. What Kaplan does not note is that this paper was more a statement of a research program and an assessment of the literature that might provide frameworks for thinking through the supposed connections between environmental change and political violence than it was a final report on research findings. In earlier formulations such matters were usually linked to questions of development rather than explicitly to matters of security. They became a major theme in the World Commission on Environment and Development's (WCED) 1987 report, *Our Common Future*, in which it was asserted that "[e]nvironmental stress is both a cause and an effect of political tension and military conflict." This widely cited report also argued that even if causative linkages are far from being simple or well understood, conflicts seen to be linked to the environment "are likely to increase as these resources become scarcer and competition for them increases."[1]

In the early and mid-1990s, some researchers argued that such statements were not proven and that the parameters of such linkages were not well understood even in situations where they might hold.[2] Critics have also charged that security agencies, and the military in particular, play a major part in environmental damage in many parts of the world.[3] The scholarly debates have addressed empirical research and policy analysis, looking at both the search for linkages between conflict and environmental degradation and the policies likely to take any such linkages for granted if only because the environment is specified as a national security concern.

As will be elucidated below, a discussion of the links between environmental change and conflict can prove difficult if at least some of the implicit conceptual assumptions related to the debate about broadening security are not made explicit. Such matters are unavoidable in making connections between security and environment and specifically in exploring whether Kaplan's dark vision of environmentally induced chaos and violence has any basis in empirically demonstrable social relationships in particular places. The methodological and substantive scholarly literature on the links between environmental degradation and conflict is thus the topic for this chapter. Rather than review the literature comprehensively, it deals with the substantive methodological debates and the findings of the most significant research efforts. The related question, which geopolitical frameworks are appropriate for considering the larger contextualization of this literature within the processes of ecopolitical change, is discussed in chapter 4.

ENVIRONMENT AND CONFLICT: METHODOLOGICAL DEBATES

While leaving aside the related concern of damage caused by military activities, Marc Levy suggested, in a widely cited critique of early 1990s research on environmental security matters, that it can be a useful initial heuristic device to divide environmental security issues into three categories: existential, physical, and political; and to test them against the security backdrop of a single state.[4] As a general principle, it can be argued that existential threats stemming from environmental destruction make humanity generally insecure.[5] Levy considered existential versions of the link between security and environment to be so undertheorized as to be nearly analytically useless, and hence the notion of security becomes so diffuse that it loses any meaning relative to the vital interests of states, especially so in the case most frequently discussed in extant literature at that time—the national interest of the United States.

If all environmental damage was a security threat, then conventional policy formulation would be impossible because neither damage nor responses would be calculable and the priorities for policy action would be impossible to establish. The counterargument is often made, however, that numerous facets of environmental damage have serious implications for many societies, and hence demand urgent attention.[6] But this argument is not related to the narrow sense of security as a matter for states; rather it is concerned with broader matters of human security. This adds the complicating matter of governance, broadly understood, and the important point that states often render their populations insecure in many ways despite, or perhaps because of, their supposed rationale as security providers.[7]

In the case of the second type of threat—direct physical damage from environmental threats to any nation in general and to the United States in particular—stratospheric ozone depletion and medium-term global climate change provide good examples. Ozone depletion is an obvious direct threat to human health. Disruptions caused by climate change may also cause agricultural and infrastructure damage due to storms, droughts, and flooding. A broad interpretation of national interest suggests that these hazards ought to be considered under an overarching notion of security. But the practice of dealing with environmental problems ironically suggests that this may not be the most appropriate policy response. This is clear in the argument that—at least in the case of ozone depletion—the widely supported 1987 Montreal protocol on limiting the use of ozone-depleting substances, and its subsequent updates, gained approval in the United States precisely because the negotiations were not treated as a matter of high politics or as a security concern.[8] Following this logic— where flexible arrangements and customized responses were negotiated to accommodate numerous widely differing interests—the use of a top-down environmental policy framework of precise rules and strict monitoring and enforcement by an international security agency has little chance of providing the flexible responses needed to counter long-term dangers presented by climate change and other possible future dangers.

The third danger to the West and to the United States is from indirect political disruptions caused by environmentally induced political conflict in economically poor states. Much of the empirical literature on environmental degradation and conflict has focused on the likely parameters of migration and the causal mechanisms that lead to the breakdown of societies and to violence and displacement. Problems of environmental refugees, collapsing states, and migration issues have all been investigated in various ways in both the popular press and scholarly writings as threats to the states of the North. Political instabilities, induced by direct conflicts over resource uses and by indirect problems generated by ensuing waves of refugees and migrants, will, so the argument made by Kaplan and others goes, cause spillover effects in the affluent Northern states that will then be obligated to provide a security response. Skeptics, however, have challenged the logic of these arguments, suggesting that while these are matters of concern, the formulations of cause and effect and policy responses are not adequately conceptualized.[9]

One method of dealing with the conceptual difficulties and the fuzziness of the term "security" has been to avoid using it as an analytical category and to focus instead on what may link environmental degradation to overt organized violent conflict. This is the approach undertaken by Homer-Dixon and his research teams in a number of projects in the 1990s. Although evidence obtained from such research has suggested that there are, indeed, some clear linkages between environmental degradation and conflict, the case studies are complicated and subject to multiple interpretations.[10] Scarcity of environmental resources is a factor in some conflicts; forced migrations sometimes lead to identity conflicts as refugee claims clash with those of local residents. Relative scarcities, which encourage elites to seize control of environmental resources, often further aggravate the marginalization of poor populations. Research on China and Indonesia, focusing on state capacity in the face of likely future environmental constraints, suggests that the links between environmental degradation and civil conflict are far from simple.

In some ways, such debates divulge much more about underlying

prior assumptions than about the plight of those bearing the consequences of environmental change. Thus, their significance depends in
part on whether or not environmental degradation is widely understood as a problem for security and on whether and why the premise
needs empirical validation prior to policy initiatives.[11] If environmental matters are already recognized to be integral to contemporary
intrastate violence, which is the argument suggested by the WCED,
and one takes for granted that environment is a cause for political
strife, then detailed research into case studies can appear less useful
than research into policy options and regional responses. It can also
be argued that whether they have spillover effects or not, the very
fact that people are being killed in some states is—in and of itself—
an important matter for scholarly investigation and a security concern, when security is understood to include broad concerns beyond
the mere safeguarding of political order in the modern West.[12] The
assumption of how security should be conceptualized is unavoidable
in this discussion.

Levy also argued that this leads to the related questions of whether
a research design that explicitly focuses on the environment is to be
preferred to one concentrating on regional conflict portraying the
environment as a causal factor, and whether some intervening variables, between environment and conflict, may not prove more important than the environment as determinants and predictors of likely
conflict.[13] If violence is a domestic issue for states and if environmental factors contribute to it, then environmental insecurity—although
of concern—may not be so directly related to regional conflict. Intervening variables may prove more helpful in predicting and explaining
the causes of violence, given the variety of political and sociological
conditions that can affect the circumstances in which conflict may
occur. The literature of the mid-1990s offered little on the causative
role that intervening variables might play in circumstances of environmental degradation. Missing still was the suggestion that these political, economic, and sociological intervening variables might in fact be
simultaneously causing violence and environmental degradation.

If the intervening variables are a crucial part of the process, then
the causes of both degradation and violence may be understood in a
rather different fashion. Such analysis would suggest that the causes of
environmental degradation are more intimately connected to conflict
and insecurity and that the larger concern is with political activities

engendered by the processes that are causing degradation. Likewise such a political ecology analysis might suggest that concerns about intergroup violence, which is the traditional focus of security analysts, may be somehow occluding the significant political actions of social movements that resist degradation and challenge the processes of enclosure and dispossession that are part of the development strategies of many states. Emphasizing this dimension of the environment-conflict nexus leads to a different problematique, one that connects up with larger questions of political economy, which will be discussed in chapter 5.

Such matters did not appear in most of the security and conflict debates on the appropriate methods and priorities for empirical research into the links between environmental degradation and conflict in the early stages of research in the 1990s. More recent research has begun to flesh out these themes, in what has sometimes been called the third generation of environmental security research, which gets beyond the early discussions of conceptual matters dealing with environment and second-generation research that established plausible linkages between environmental change and conflict.[14]

The remainder of this chapter outlines the two most comprehensive research projects on environmental change and violent conflict, Homer-Dixon's work and the ENCOP (Environment and Conflicts Project) in Europe, summarizes their findings, and draws some very tentative conclusions on their geopolitical implications. Other research is obviously relevant, but the conclusions of these two projects are very similar, and while many authors may object to some facets of the results or their conceptual structures, other research has not substantially challenged the empirical findings discussed here. These projects are also significant in that they provide a substantial part of the conceptual structure for the NATO discussion of environment and security.[15] Homer-Dixon's framework in particular is influential in the discussion of the "pivotal states" strategy proposals for U.S. foreign policy drafted by Paul Kennedy and his Yale University collaborators.[16]

ENVIRONMENT, SCARCITY, AND VIOLENCE

In an attempt to render the research task practical and relatively rigorous, Homer-Dixon's approach focused on violence both between and within states. This excludes many other forms of political dispute. Another important dimension is his attempt to show fairly direct causal links between environmental scarcity, to use his term, and overt conflict. These stipulations both narrow the empirical range of phenomena for study and simultaneously bracket questions of security by avoiding much conceptual debate. The definitional questions about scarcity are immensely complex, even if matters of conflict can be resolved by emphasizing overt and relatively large-scale collective violence. It is important to note, however, that the assumption of scarcity is basic in the discussion that follows, although the causes of conditions of scarcity in particular cases are not simple.

Homer-Dixon's recent major book on the questions of scarcity, environment, and violence summarizes the research results from the projects that he directed on "Environmental Change and Acute Conflict" (ECACP) (1990–1993) and on "Environment, Population, and Security" (1994–1996). Subsequently, he has also worked on the question of state capacity and the ability of states to deal with environmentally induced difficulties. He starts by classifying five categories of dispute that could plausibly be caused by environmental scarcity.[17] First are disputes arising from site-specific concerns such as logging, pollution from a factory, or dam construction. Second are ethnic clashes caused by migration and social cleavages caused by environmental scarcity. Third are matters of civil strife caused by environmental scarcity that affects economic activity, livelihood, elite behavior, and state responses. Fourth are scarcity-driven interstate wars, possibly over water supplies. Fifth are large-scale North-South conflicts related to global problems of climate change, ozone depletion, biodiversity, and fishing. Disputes in the first and fifth categories are unlikely to lead to organized violence, and the fourth category, interstate warfare over water, is, Homer-Dixon suggests, unlikely. Hence the emphasis in this research is on the second and third themes: ethnic clashes, migration, and social cleavages, and state and elite actions, all in the face of scarcity.

Homer-Dixon's research is extensive. His projects in the 1990s generated two books, numerous articles, working papers, and reports. What is of interest here is only the broad outlines of the research findings and the key finding that supports the contention in the World Commission on Environment and Development (WCED) report that environmental scarcity is a factor in causing violence. Much of Homer-Dixon's research is thus a matter of process tracing, looking at how plausible links between one factor and another can be constructed. It requires drawing on numerous disciplinary specializations and modes of analysis. Above all else, his research has use in showing the complexity of these links and revealing just how little hard evidence there was in the 1980s and early 1990s to support sweeping statements that environmental degradation leads necessarily to conflict. His research reveals, he claims, the necessity of adding environmental considerations to discussions of the causes of conflict. He does not argue that there is any obvious or predictable relationship between environmental factors and conflict, only that, in a number of cases large enough to be worthy of serious scholarly and policy attention, environment can be shown to be an important factor in the multiple causes of civil violence.

Key to all this is the matter of environmental scarcity; its definition is crucial to the argument that degradation leads to conflict. Homer-Dixon suggests that scarcity is a matter of three interlinked factors he calls supply-induced, demand-induced, and structural scarcities.[18] Supply shrinkages happen when renewable resources are degraded or in such circumstances as a drought. Demand increases when population increases. Structural scarcity is induced by elite attempts to gain a larger proportion of a resource, leaving some of the population without adequate access. These phenomena interact in ways that can lead to resource capture where "a fall in the quality or quantity of a renewable resource interacts with population growth to encourage powerful groups within a society to shift resource distribution in their favor."[19] When this is connected to population growth that results in people moving to marginal lands that are environmentally vulnerable, the result is what Homer-Dixon calls, following J. Leonard, ecological marginalization.[20] It is necessary to emphasize the importance of contextual circumstances in those situations where scarcity is turned into violence. Homer-Dixon argues that environmental scarcity might be best understood as an underlying stressor of many social systems rather than as a direct cause of conflict.[21]

The ability of states, as the most important social institutions in most places, to respond to these processes is obviously key to understanding where social breakdown and violence occur. Declining state capacity, in the Homer-Dixon framework, is related to increasing environmental scarcity in at least four ways. Increased financial demands on the state for infrastructure is one factor. Second is the problem of dealing with elite demands for assistance or law changing, in part because, third, predatory behavior by elites may lead to defensive reactions by weaker groups. Fourth, a general reduction in economic activity reduces state revenue and fiscal flexibility.

From his model Homer-Dixon classifies the resulting conflicts into three types. First are simple scarcity conflicts, which have traditionally been used to explain wars as varied as the Gulf War and the First World War. However, Homer-Dixon suggests that there is little recent evidence of international conflict over environmental resources. Agricultural land, forests, river water, and fish are apparently unlikely to generate wars in the future due to their relatively indirect contribution to national war efforts and state power. Homer-Dixon notes that forests and supplies of timber for ship construction were an important factor in the past, but this concern disappeared with the emergence of coal, oil, and steel as the key materials for war machines.[22] Water wars are a matter of internal conflict rather than international warfare, partly because there are only a few situations of strong downstream states likely to be willing to go to war with upstream states over supply issues, and partly because water disputes are often complicated jurisdictional issues with many players who do not easily polarize into warring coalitions.

Second are group identity conflicts, which occur where migration conflicts are generated by, or aggravate access to, environmental resources in the urban areas where recent migrants compete for limited supplies. These migrants are often weak and relatively ill organized, but conflict can arise as a result of many factors, as the confused patterns of violence in postapartheid South Africa suggest.[23] These conflicts can have transboundary effects, too, for example, migrants from Bangladesh are involved in conflicts with people across the border in India.[24]

Insurgencies are the third category; the situation in Chiapas is a classic example of a direct challenge to the state. Here organized opposition to an oppressive state has taken the form of grassroots organization, political mobilization, and at least some military action. He might have added that the Zapatista struggle shows much ingenuity in the cause, in particular in the use of international coalitions of support mobilized in part through the use of the Internet.[25]

Homer-Dixon's policy suggestions are not unique but stress that action is necessary to prevent a downward spiral of scarcity-induced social changes that will eventually lead to violence. Of significance for the larger discussions of security, he suggests that most of the actions needed are not of the sort that military institutions are well equipped to handle.[26] His analysis points to the need for ingenuity in the face of social change and the stresses that will come from environmental scarcity. Nonetheless, the danger of a number of the larger poor states in the South either imploding or becoming dangerously authoritarian, or "hard states," should not be ignored, because of the likely consequences for the populations of those states and for the likely international repercussions of these developments.[27]

Here Homer-Dixon's analysis linked up with the larger concerns in American foreign policy circles about the future fate of China and the possibility of it following the same path as the Soviet Union to dissolution. Research projects under Homer-Dixon's direction included analyses of the capacities of the largest states likely to suffer from environmentally induced conflict.[28] Analyses of China in the 1990s were concerned with shortages of resources to feed the population and to fuel its rapidly expanding economy. The growing contrast between prosperity on the coast and poverty in the interior raised speculations that environmental stresses might trigger the violent collapse of the state and perhaps even a reversion to the era of the warlords, this time with modern weaponry and nuclear devices involved. While such alarmist scenarios have frequently been dismissed, the future of China remains a high-profile issue in the Woodrow Wilson Center's research into environmental change and security.[29] But much of this discourse remains caught in the scripts of geopolitical containment, in terms of reducing the threat of internal disruptions to external relations or matters of American leverage over China, in a replay of the cold war discourses of food power as a policy tool for Washington's use.[30]

The focal point for Homer-Dixon's analysis is the ability of societies to adapt to rapid change and the environmental degradation that goes with development. State adaptation per se is not necessarily the most important theme, although it clearly matters, not only in the case of China. Indonesia is another case of concern in this framework, one that was especially prominent as the 1990s came to a close and political and economic turmoil coincided with widespread forest fires.[31] Homer-Dixon emphasizes social and technical ingenuity as key to this.[32] Societies that have the ability to generate new economic structures and can make technological and social innovations will avoid the downward spiral to poverty and violence. But

> If a country loses the race—if, in other words, a gap develops between ingenuity requirement and supply—social dissatisfaction will rise, with increasing stress on marginal groups, including those in ecologically fragile rural areas and urban squatter settlements. . . . Violence further erodes the society's capacity to supply ingenuity, especially by causing human and financial capital to flee. Countries with a severe ingenuity gap therefore risk entering a downward and self-reinforcing spiral of crisis and decay.[33]

THE ENVIRONMENT AND CONFLICTS PROJECT

The ENCOP team, based in Switzerland under Gunther Baechler's overall direction, came to broadly similar conclusions about how the conflicts in the South could be understood and categorized. However, their mode of inquiry proceeded somewhat differently. Baechler's detailed explication of the ENCOP analysis draws on a large series of case studies of conflict conducted by project researchers.[34] Using aggregate national statistics and country comparisons, he correlates war proneness in the mid-1990s with the United Nations human development indicators (HDIs), finding, perhaps not very surprisingly, that those states with the lowest HDIs had the highest proneness to warfare. Southern Africa is especially war prone and has states with the lowest HDIs.[35]

Baechler's statistical calculations further suggest that the states with the lowest HDIs also have the highest percentages of their populations involved in agricultural activities. Rural poverty is a major problem, though the populations of even the poorest states are gradually urbanizing. Low amounts of cropland per capita combined with low commercial energy per capita consumption indicates states that are highly war prone. Urbanized developed states are not directly dependent on natural capital as most economic activity occurs in non-agricultural sectors. The overarching geographical theme in these discussions of environment and conflict is the presence of most conflict that can be linked to environment in the poorer areas of the world, in the South. In Baechler's terms, "The sheer coincidence of underdevelopment, transformation of landscape, and violent conflicts or wars in many regions of the South and East leads to the supposition that analysts are not confronted with an isolated phenomenon in one specific area alone. Many local or regional forms of conflict conform to a general basic pattern."[36]

The definition of an environmental conflict that the ENCOP research team used emphasized that these were traditional-style conflicts triggered by environmental matters.[37] Environmental degradation can consist of the overuse of renewable resources, pollution, or the impoverishment of living space. Although excluding nonrenewable resources and minerals, this definition does include activities of mining and dam building that seriously degrade local spaces. From the ENCOP analyses, "it is clear that neither apocalyptic scenarios of environmental catastrophes nor alarmist prognoses of world environmental wars are tenable."[38] This concurs with Homer-Dixon's conclusions that environmentally induced conflicts are mostly diffuse and subnational.

The ENCOP analysis developed a comprehensive hypothesis that "Violent conflicts triggered by the environment due to degradation of renewable resources (water, land, forest, vegetation) generally manifest themselves in socioeconomic crisis regions of developing and transitional societies if and when social fault lines can be manipulated by actors in a way that social, ethnic, political, and international power struggles occur." That is, violence is likely to occur in specific places where groups that are discriminated against are in situations of environmental degradation. The concept of environmental discrimination is introduced to make this point clear: "environmental discrimination occurs when distinct actors—based on their international position and/or their social, ethnic, linguistic, religious, or regional identity—experience inequality through systematically restricted access to natural capital (productive renewable resources) relative to other actors."[39]

The geographical situations where such conflicts are possible are in arid and semiarid areas, mountain areas with highland-lowland interactions, river basins crossing state boundaries, dam-building and mining areas, tropical forest regions, and poverty clusters in large metropoles. But while these are obviously geographically defined locales, environmental determinism is rejected in this analysis: "passing the threshold of violence definitely depends on sociopolitical factors and not on the degree of environmental degradation as such." The overarching ENCOP conclusion suggests that violence occurs when some combination of five key situations occur:[40]

1. Inevitable environmental conditions where a group is dependent on degraded resources for which there is no substitute
2. Scarcity of regulatory mechanisms and poor state performance
3. Instrumentalizing the environment by a dominant group so that environmental discrimination becomes an ideological issue of group identification
4. Alliance-building opportunities
5. Spillovers from historic conflicts

In summary, Baechler concludes that "[a]cute conflicts do not arise along the great fault line between North and South, but rather where climate change contributes to the collapse of local rural structures and regional political authorities."[41] Although proceeding in different ways,

> There is only one finding where ENCOP results differ from ECACP. While Homer-Dixon suggests a linearity between large population movements and group identity conflicts, ENCOP suggests that migration is linked to different kinds of conflicts: socioeconomic conflicts between highland and lowland producers, conflicts between rural and urban dwellers, and conflicts between rural producers and central states' forces. Migration also causes conflicts within one and the same ethnic group, which may be divided by geographical or national boundaries.[42]

GEOGRAPHIES OF INSECURITY

The ENCOP analysis suggests, perhaps more clearly than Homer-Dixon's presentation, that conflicts have very specific geographies, most often occurring where environmental degradation coincides with political divides and unequal access to resources:

> There is ample empirical evidence for the basic assumption that maldevelopment induces both environmental transformation and violent conflicts. The three issues must be seen in a triangular constellation, since intractable poverty and conflict-prone environmental transformation become more and more closely intertwined in particular ecogeographic zones. These zones encompass both dry land and sensitive mountain arenas with large rural populations being increasingly dependent on relatively shrinking natural capital in general and scarce as well as degraded renewable resources in particular.[43]

The crucial point in the ENCOP case studies is that not all situations of environmental distress turn into violent conflicts. "[T]he reasons why these conflicts became violent are found in the sociopolitical network in terms of poor state performance, repressive regimes, lack of regulatory mechanisms, and unavoidable situations."[44] The last point in this list suggests that in some situations people simply have no options other than to fight.

Mountain areas and arid plains are of particular concern in Baechler's analysis. Arid plains have traditionally been home to nomadic herders. As the African Sahel shows, the penetration of modern states and commercial economies into arid lands has often disrupted migration routes, attempted to settle nomadic populations around new wells and town sites, and in the process of extending the rule of state law partly superceded traditional dispute resolution mechanisms among pastoralists, frequently leaving these arid plains poverty-stricken and environmentally degraded, as well as politically unstable.[45]

Mountain areas are often remote, Baechler argues, the refuge of old cultures where modern, more powerful cultures have only sporadically penetrated. Cultural divisions often follow topographic boundaries. With the notable exception of the Ethiopian highlands where settlement has flourished in the hospitable environments there, most mountain areas are relatively unproductive, and with their steep gradients and high precipitation levels are supposedly vulnerable to degradation.[46] They are frequently frontier zones and the source of outward migration to the more prosperous lowlands. Complex historical patterns of conflict have often involved disputes related to transhumance patterns when pasture, either in the highlands or winter grazing in lowland areas, became overused.

The ENCOP authors and Homer-Dixon make it clear that those who suffer in these disputes are the local populations dependent on environmental resources for practical survival. But how these conclusions are to be interpreted in geopolitical terms or used to provide policy advice or useful contributions to the discussion of reformulating security remains less developed. Baechler is prepared to offer some suggestions on this score, but Homer-Dixon prefers to shift attention to the provision of ingenuity. Specifically, Baechler suggests that all of these processes might be seen in a larger framework, drawing from Karl Polanyi's discussion of the "great transformation" from feudal and agricultural to industrial society. More specifically, Baechler suggests that Polanyi focused on the socioeconomic side of modernization in the North, but discussions of environmental conflicts should focus on the "global socioecological side of adverse impact of modernization and development in the South."[47] Although it is not a major theme in his book, it is worth noting that Polanyi clearly understood that this was a long-term process going back at least to the dramatic changes in the second half of the eighteenth century in England:

> The mobilization of the produce of the land was extended from the neighboring countryside to tropical and subtropical regions—the industrial-agricultural division of labor was applied to the planet. As a result, peoples of distant zones were drawn into the vortex of change, the origins of which were obscure to them, while the European nations became dependent for their everyday activities upon a not yet ensured integration of the life of mankind. With free trade the new and tremendous hazards of planetary interdependence sprang into being.[48]

Baechler understands the violence and maldevelopment as in part a matter of an imperfect penetration of modernization that is transforming the landscapes of Southern rural areas. The Sahel region, Baechler suggests, epitomizes these processes and is home to much of the post–cold war violence in Africa. Thus current environmental conflicts are part of the continuation of the "great transformation," albeit with a focus now on natural resources that play a part in the violence of the processes. The implications in this formulation are very considerable. Similar arguments occur in Homer-Dixon's discussions of transformation and the dangers of lack of ingenuity, but Baechler's analysis more explicitly links these to both preexisting historical patterns of conflict and political divisions at the small scale, as well as to the large-scale connections.

Understood in these terms, much of the environmental conflict going on in the 1990s could be understood in core-periphery terms if the indirect effects of modernization are also seen as part of the processes that interconnect with environmental degradation. In the ENCOP case studies, the largest category of conflicts is the center-periphery one where state elites cause all sorts of insecurities for their own populations.[49] Where states are not clearly acting in the interests of their populations, the assumption usually is that they will do so when their level of economic development rises high enough. This may require significant infrastructure building and the conversion of subsistence economies to modern commercial ones.

This crucial assumption is vital to understanding the processes rendering many populations insecure as they go through the local version of the "great transformation." In the process of development, from the Amazon to the project to build dams on the Narmada River in India, large numbers of people may be displaced and enter into direct conflict with state forces and corporations.[50] Modernization is a process that often involves conflict. Development itself can be understood as a violent process where people are uprooted; lands are flooded for dams, covered by roads and pipelines, or converted to commercial use; while waterways bear the brunt of pollution from mines and agricultural chemicals.[51]

Where states are the direct providers of insecurity, as is often the case with these large infrastructure projects, the provision of security cannot be separated from the political questions of who is thereby being made insecure. Environmental displacements leading to conflicts of identity are one possible cause of violence. But as the ENCOP studies of center-periphery conflicts suggest, serious insecurity is caused by state-supported or state-imposed "developments" that engender resistance and opposition from people who find themselves banished, while environmental resources are appropriated and access to them withdrawn via the commercial-industrial arrangements of cattle ranches, clear-cut logging, or agricultural plantations. In Homer-Dixon's terms, resource capture and environmental marginalization might be understood here as integral to state building.

The complexity of the interconnected social processes demands careful analysis because not all inhabitants of an ecosystem cause similar damage to natural processes, and the perpetrators of disruption often are not the immediate victims. Endogenous scarcity-induced conflicts are thus likely to be less violent than those caused by obviously external forces, because, to put the matter in Lothar Brock's terms:

> The situation is different where interventions in local resource use are external and clearly identifiable, as through dam building, mining, commercial logging of forests or the displacement of small farmers by large landowners. As here the responsible parties are clear (or perceived as being so), there is a greater danger of violence in such conflictual constellations than there is in use conflicts arising from endogenous scarcity. The violent resource use conflicts in the Brazilian Amazon are an example of this. Here the subsistence possibilities of

small farmers are attenuated by external users (expanding large land-owners, mining). However the small farmers, for their part, are in conflict as external users with the indigenous peoples of the forests (and the rubber tappers). It is thus expedient to distinguish between poverty related and commercially generated exogenous environmental scarcity.[52]

NATO SYNDROMES

In reviewing the literature on environmental factors and their relationships with conflict, the authors of the 1999 NATO study on environment and security drew heavily on both Homer-Dixon's framework and the ENCOP research.[53] They emphasized the importance of "environmental stress" as a factor in possible conflict and explored the avenues for linking environmental stress with other social, demographic, and political factors. Once again the sheer complexity of the possible interlinkages becomes clear in what are here termed nonlinear complex systems. So too does the need for interdisciplinary research if such systems are to be subjected to systematic study by social scientists.

In developing their proposals for a system of indicators that might allow for the monitoring of situations and provide early warning of potential violent conflict, the NATO authors also move beyond the Toronto and ENCOP frameworks to draw from work by German researchers on syndromes of change in specific places.[54] Once again using a medical language of stress and syndromes, monitoring and interventions, this approach has the advantage of focusing on the specifics of social circumstances to disaggregate environmental factors and hence clarify causal mechanisms. Thus soil erosion in the United States caused by industrial farming is part of a dust bowl syndrome and not linked directly to the erosion caused by poverty-related overuse in Africa where it becomes part of a Sahel syndrome.

Three broad categories of utilization, development, and sink syndromes deal with particular combinations of potential environmental difficulties that might lead to conflict, although the syndrome of environmental destruction as a direct result of war and military action obviously makes the relationships explicit and direct, albeit in a reverse causation. What is implied in these are interconnections with the larger global polity, but the preliminary analysis here is not explicitly linked to either global developments or violence. However,

the overexploitation syndrome, where natural systems such as oceanic fish are depleted, and the smokestack syndrome, degradation
through large-scale diffusion of long-lived substances, might be explicitly understood as all-encompassing problems. The focus in the
Aral Sea syndrome is on the international consequences of large-
scale technological developments and the possibilities of such things
as water wars.

These discussions inevitably link back to the questions of sustainability and the policy prescriptions that might alleviate the utilization
syndromes without either introducing development syndromes or aggravating sink syndromes. In particular, the question of indicators
only emphasizes how complex the questions of interconnections are
in discussing causation. But clearly this line of reasoning suggests the
importance of identifying threshold values for specific variables, which
will obviously be different depending on the syndrome in question in
the particular region under consideration. Despite the number of
indicators already in existence, finding ones that will allow global
monitoring of environmentally induced conflict potential obviously
requires considerably more conceptual innovation and practical research. The political utility of such indicators might be obvious to
NATO officials anxious to monitor the trends that cause political turmoil, but what they might mean to the populations so monitored is
much less clear. Of course, the security that is of most concern here is
the political stability of the world system. The possibility that it might
be engendering insecurities is not entirely dismissed in the discussion
of the various syndromes of change, but a more dramatic rethinking
of security is not countenanced in this discussion.

ENCOP AND ENVIRONMENTAL SECURITY

Baechler ends his discussion of the ENCOP model and the Rwanda
case with a discussion of the implications of the analysis of environmental conflicts for rethinking environmental security. Noting that
many of these difficulties are discussed in parallel in the environmental security and the sustainable development discourses, Baechler suggests linking the two discussions together. "Sustainable development
is associated with vulnerability and its reduction, while environmental security is related to threats and their minimization; both corre-
late positively." While development dilemmas are more important
than security dilemmas in many aspects of the transformation, threats

(security) and vulnerability (development) are often closely linked. Both need attention in policy measures to facilitate transformation. "Sustainable development focuses on the dilemma between economic growth and a way of resource use that respects the regeneration rate of renewable resources."[55]

In parallel, "[e]nvironmental security is focused on the dilemma between securing and preserving societal and political institutions and leaving room for dynamic development."[56] But more than this:

> Environmental security is an indicator if and how unsustainable development—which increases vulnerability—threatens social and political entities within the framework of society—nature relations. A lack of ecosystem development, for instance, can seriously affect behavioral, political and legal measures and therefore provoke a perception of ecosystem insecurity. And unsustainable human development can seriously threaten livelihood security in all its major aspects.[57]

But the policy implications of these matters suggest the importance of getting out of conventional modes of both security and development thinking. Conventional answers to unconventional difficulties will often not allow the intersection of security and development to occur in a way that reduces both vulnerabilities and threats. This is implied by the NATO focus on syndromes but not clearly developed in any practical ways. It is why ingenuity is so important.

To illustrate the importance of unconventional thinking, Baechler concludes his book with an example of a solution to environmental conflict among seminomadic peoples of Borana in northern Kenya and southern Ethiopia.[58] Here a complex agreement between peoples over grazing rights, confirmed with important rituals of agreement, was complemented by a number of communal efforts to increase diversity of the economic base for the survival of all concerned. This arrangement was not complicated by the involvement of either of the states concerned, which may be an important contributing factor to success because the traditional African understanding of land as involving right of use, not absolute property rights as in modern law, is important to the solution.[59] Local understandings of the trade-offs of matters of value were employed to move discussions along, and a group of monitors, selected in common by both sides, was established to settle any disputes that arose. The crucial point is that threats of violence were removed and the possibility of changing the economic base to prevent degradation leading to conflict in the future was built into the approach. In Homer-Dixon's terms, these social innovations are clearly the practice of ingenuity, although not a matter necessarily of technical know-how or formal education. Local knowledge of environments and social systems was key to the process.

What is especially important is the recognition of the dynamics of change. Security is not about protecting a stable status quo from an external threat. It is about developing an economic system that reduces dependence on a single resource, a dynamic system that can accommodate change. This does not fit easily into traditional understandings of defense or national security, which is, of course, the whole point of trying to rethink security. The "Borana solution" does fit into understandings of common security and the need to evade the difficulties of escalating threats in a situation of a security dilemma. The novelty of such understandings suggests that environmental security cannot be understood in terms of state defense policies, either as military preparation or territorial surveillance. Nor can it be linked to static models of unchanging environments in some traditional national park metaphors or to the assumptions of simple equilibrium models of sustainable yield, where one of the crucial factors is the fact that environments are being changed by the patterns of resource usage. Obviously ecological matters need careful attention, but the interesting point about all these analyses of environmental security is the linkage to the processes of contemporary transformation.

The larger question is one of the international connections that cross boundaries, specifically how much internal change in one state is driven by actions in another. For conceptual reasons these links are an important corrective to treating states as black boxes and environmental causes as endogenously caused. For policy reasons these interconnections are precisely where foreign and trading policies might make a contribution to changing conflict dynamics within states in the South. Insofar as environmental security can be considered to be a global matter, these interconnections are significant.[60]

The historical dimensions of the "great transformation" are also directly relevant to discussions of environmental security. While the potential conflicts in the areas that the ENCOP researchers identify are important, there are numerous historical precursors. The "great transformation" involved a dramatic series of changes to European and

North American societies but also entailed environmental changes close to home and further afield. The growing trading system, as well as imperial conquests, increasingly provided Europeans and Americans with commodities from distant places. But these exports inevitably involved the displacement of peoples and economic systems elsewhere. This crucial geographical point has not always been clear in the environmental security discussion. It is completely absent from Robert Kaplan's considerations, but it needs incorporation directly into the analysis to explain the processes of dislocation that may lead to local violence. The following chapters show that it also needs to be incorporated to link the discussion of environmental conflict to both the larger concerns of global politics and to the possibilities of rethinking security priorities in many places.

4

Geopolitics and History: Contexts of Change

Representations of space in the social sciences are remarkably dependent on images of break, rupture and disjunction. The distinctiveness of societies, nations, and cultures is predicated on a seemingly unproblematic division of space, on the fact that they occupy "naturally" discontinuous spaces. The premise of discontinuity forms the starting point from which to theorize contact, conflict and contradiction between cultures and societies.

AKHIL GUPTA AND JAMES FERGUSON, "BEYOND CULTURE"

Contemporary global understandings remain attuned to historical narratives that naturalize a particular, territorially oriented view of sovereignty, reinforce it with a political economy story that disparages pre-commercial systems of livelihood and exchange, and substitutes myths of evolutionary development for histories of violent confrontation and usurpation.

MICHAEL SHAPIRO, *VIOLENT CARTOGRAPHIES*

PLACING SECURITY

Robert Kaplan raised the specter of political chaos induced by a Malthusian mismatch between population growth and resource inadequacies. As chapter 2 made clear, using the political failures of West Africa as his suggested model of the future, this argument constructed the world in imaginary spaces of chaos and anarchy contrasted to the affluence of the developed world. The danger lies, he argued in 1994, in the spillover of chaos from the zones of turmoil into the zones of peace.[1] The most obvious dangers are in the forms of drug trading, disease, and large-scale transnational crime that will merge with new forms of warfare and terrorism. This division of the world into separate geopolitical regions is important to both Kaplan's argument and much of the rest of the popular literature on environmental security where the underdeveloped global South is portrayed as the source of instabilities that threaten the North.

The focus on geopolitical dangers rarely examines the specific causes of degradation in any detail. Often, in parallel with Kaplan, the environment is reduced to an exogenous factor or an independent causal variable in analyses of conflict. This approach is usually in danger of replicating the faults of environmental determinism in attributing simple causal power to natural environments in human affairs while disregarding the specificities of human institutions in particular places.[2] As subsequent chapters will explicate, it also frequently overlooks the important cross-boundary patterns in the flows of resources and the politics involved in the dispossession of peoples to facilitate resource extraction and the expansion of commercial agriculture. But before turning to consider indigenous peoples and their politics, this chapter extends the analysis of new thinking in security by looking more closely at the geopolitical assumptions inherent in conventional security analyses of environmental themes.

Given current criticisms of the geographical and spatial premises of international relations thinking, just what are the appropriate geographical frameworks for investigating matters of environmental change and conflict? This chapter looks briefly at two arguments in the current international relations literature on environmental matters in order to investigate their geopolitical presuppositions: first, the environmental dimensions of the democratic peace; and second, the matter of providing aid to states in the South to deal with environmental difficulties. Showing how the spatial premises of these arguments limit the analysis in a way that fails to deal with the processes causing degradation by inadequately considering the appropriate scale, it also suggests the need to think about the temporal dimensions of the processes involved. The post–cold war claims that environment is a new consideration for security analysis also raise the question of the appropriate historical context in which to think about matters of environment and conflict. This is the subject of the latter part of this chapter.

ENVIRONMENTALLY BENIGN DEMOCRATIC PEACE?

While much of the literature on environmental security is pessimistic, if not downright alarmist, at least in its popular articulations, some other contemporary international relations literature on warfare is much more optimistic about the future. Contemporary discussions of the "democratic peace" suggest that advanced industrial democracies do not fight each other and that major wars are obsolete in the nuclear age.[3] Coupled to the optimism that the end of the cold war will lead to the expansion of the number of democratic states in the world, this suggests that the future may avoid the major violent interstate confrontations of the past.

When the "democratic peace" argument is linked to considerations of the supposedly superior records of industrial democracies in environmental protection, it can be suggested that there is considerable hope for dealing with environmental security questions.[4] It seems that environmental problems occur more frequently in parts of the world that have authoritarian or communist regimes, so democratization should lead to environmental improvement by enhancing political accountability. Here is the possibility of a virtuous circle: democracies do not usually fight each other, nor do they produce environmental destruction that might, according to the "environmental degradation leading to conflict" argument, lead to instability and violence.

But there are a number of geopolitical arguments related to this position that need to be taken seriously in thinking about the formulation of an "environmentally benign democratic peace." The first apparently strengthens the case in a slightly ironic way. While democracies are, according to recent scholarship, much less likely to fight each other than authoritarian regimes, some scholars have suggested that the historical record of the twentieth century suggests that advanced socialist states are even less likely to fight.[5] Whatever the empirical merits of this case, given the environmental record of the Soviet Union, clearly lack of bellicosity is not necessarily related to environmental performance. Industrialization is not the issue in these formulations, democracy is.

The second geopolitical argument about the tendencies of democracies to be both peaceful and environmentally responsible suggests that the model is missing a number of crucial dimensions. In the period of the cold war, democracies often exported their violence, getting involved in some nasty wars in the South, but the argument about the democratic peace can be saved because the democracies did not fight each other, even by proxy. The tendency to get involved in violence in what is now termed the South is important in understanding the concept of democratic peace in the current geopolitical context. The model of a zone of peace is premised on a zone of turmoil outside.

Third, by relying on such a formulation of the terms of understanding of the environmental security problematique, the environmentally benign democratic peace argument does not include the point that many of the resources for Northern consumption are made available at the cost of degradation in the South. This once again raises the questions of state-territorial assumptions underlying the geopolitical units made into the objects of analysis in the discussion of the environmentally benign democratic peace. It also suggests that environment has to be understood in terms of more than "tailpipe" emissions cleanup or aesthetic tropes of hygiene; total resource throughputs in ecosystems must be included.

If the Soviet Union was, as the center of a geopolitical bloc, both an advanced industrial socialist state and an environmentally destructive one, might this be because, at least in part, its resource extractions and its dirty industries were within its borders? In comparison, the advanced democratic states have often acted in ways that effectively distance themselves from the worst sources of pollution and degradation. If the non-Soviet world is understood as one geopolitical bloc, the comparison might suggest that the environmentally benign peace is convincing only because the hinterlands that are despoiled are substantially removed from the democratic states by the global pattern of resource trading and "pollution exports."

ENVIRONMENTAL INSECURITY AND THE TERRITORIAL TRAP

As noted briefly in chapter 3 in discussing Kaplan's geopolitical imagination, the difficulties with the emergence of global arguments about environmental threats and the politics of specifying the nature and geography of the threat are reinforced when the geographical assumptions of the "territorial trap" in the domain of international relations are added to the discussion.[6] In John Agnew's terms, the territorial trap occurs as a result of three assumptions prevalent in international relations thinking. First, states have exclusive powers over their territories in terms of sovereignty. Second, domestic and foreign realms of state activity are essentially separate spheres of activity. Third, the boundaries of states define the boundaries of societies contained within those states. Combined, these three elements lead to a state-territorial understanding of the workings of power that reifies the practices of sovereign states to suggest that they are autonomous permanent entities rather than understanding them as temporary, changing, porous arrangements.

The operation of the territorial trap is especially clear in a mid-1990s scholarly volume on "environmental aid" edited by Robert Keohane and Marc Levy. Although this work is not specifically about environmental insecurity, its premises are similar to most of the key arguments in the environmental degradation leading to conflict literature and in the concern with transboundary environmental security threats. This investigation focuses on the institutions that are appropriate to channel expertise and finance to poorer states to help alleviate environmental problems there. On the very first page of the introduction to this volume the reader is invited to imagine two maps of the world, one showing the "relative severity of environmental problems," the other the distribution of "the capabilities governments have to cope with these problems."[7]

The territorial trap suggests that the variation in problems and capabilities is derived from endogenous factors within autonomous entities. These assumptions about autonomous states obscure a crucial third map of the flows of resources, which are the sources of wealth that provide governmental capabilities and which cause some key parts of the environmental degradation. But only the third imaginary geopolitical map, showing the transboundary flows of resources, suggests that the interconnections between the first two maps are an essential part of the processes that matter in understanding environmental insecurity and the causes of both degradation and government capability.

These flows of resources in the third map are obviously not the whole story, but they ought to be crucial to those in the Northern states who make policies to deal with global problems precisely because they are the part of the global picture over which the Northern geopolitical theorists of security can potentially have some immediate influence. This is because the large picture of global insecurities is cut through by an overarching irony that needs to be kept in mind, but which is obscured by Keohane's cartographic specifications. The wealthy of this world have by far the largest environmental impact on the planet and usually have the means to avoid the impacts of their actions.[8] Specifying environment only in terms of pollution or of local conditions such as water or air quality obscures the larger ecological focus on the total impact of human activities, as well as the differential impacts of these activities in various places.

Past emphases on common vulnerabilities are unlikely to hold as points of departure in future international fora. Political negotiations of global issues help reveal the limitations of conventional strategies of regime formation in the face of, among other things, the power of global corporations and other transnational actors.[9] But more important to discussions of security and international politics, the South and its new breed of policy makers now insist on discussing issues that the North's nomenclature of future global dangers downplays or ignores. In the discourses of environmental threats, chaos as a result of state failings, migration fears, or threats of disease, there is little recognition of the flows across borders that may be partly responsible for the phenomena that are now feared. In the face of specifications of these global dangers as coming from the poor and the South, any political dialogue and grand bargains over justice, development, and economic arrangements seem doomed to failure.[10]

In short, the question of the location of environmentally benign societies is once again in part a convenience of territorial state boundaries and the assumptions of states as the containers of both politics and environments. The example comparing blocs, rather than states, suggests that the geopolitical categories used in discussing environmental security are an important consideration that cannot be taken for granted. The exclusion of resource flows and other transboundary processes from state-territorial understandings of politics brings us back to the argument in chapter 3 concerning environmental discrimination and to Gunther Baechler's discussion of the disruptions caused by the expansion of modernity, processes of colonization, and the expansion of the global economy. If the spatial frameworks for discussing environmental security work to obscure the processes in motion, what about the questions of history and the appropriateness of the temporal scales of international relations?

ENVIRONMENTAL HISTORY

Much of the focus in the environmental security discussion has been on projecting current trends into the future to anticipate likely disasters and instabilities that hopefully can be avoided by adopting appropriate policies in advance or, in Kaplan's case, by being prepared for the coming chaos. This has been usefully challenged by Thomas Homer-Dixon's insistence that environmental disruptions are *already*, in part, causing political conflict. But the questions raised by investigations of environmental history have yet to be comprehensively incorporated into the discussions of either the larger questions of environment and security or the more focused scholarly investigation of the environmental degradation leading to political conflict hypothesis. They also need to be linked to discussions of security because environmental history suggests that assumptions of normal weather patterns or stable and predictable climates are not an adequate premise for discussing matters of environmental change. Change within fairly wide parameters is the long-term norm in ecological phenomena, so arguments that try to normalize particular patterns may operate to obscure natural fluctuations.

Mike Davis's harsh critique of the failures of environmental planning in the Los Angeles area and the politics of environmental threats there makes the implications of this especially clear. He suggests that cultural assumptions of a relatively stable, wet, and benign environment that English and New England settlers brought with them to Southern California might be termed the "humid fallacy." "Imaginary 'norms' and 'averages' are constantly invoked, while the weather is ceaselessly berated for its perversity" in a region that has a climate that is highly variable and where annual rainfall is only within 25 percent of the "average" 17 percent of the time.[11] Not surprisingly, as he notes, the rich homeowners who build luxury houses in the historic routes of chaparral fires through the Malibu Canyon, and elsewhere in the mountains outside Los Angeles, react with outrage and demand government disaster relief when the eventual fires sweep through their subdivisions destroying their dream houses and, if they don't drive away in time, their luxury sport utility vehicles. That there will be such fires, despite the most intense efforts at fire prevention, is inevitable in the circumstances, although not predictable in advance for any particular year. Long-term variability only assures the residents that it will burn, eventually, in the Santa Monicas.

If questions of environmental history are addressed seriously, then the geopolitical assumptions that structure the contemporary questions of the politics of global change at the largest scales may also need to be changed. Ecological history is a relatively new scholarly enterprise, but its accounts of human-induced change point to the importance of taking the long-term view and of being skeptical of contemporary claims that environmental changes offer novel political hazards. The work of historians, ecologists, geographers, paleobotanists, and scholars from a variety of other disciplinary perspectives is relevant, but synthesis is not easy.[12] What is clear is that concerns with climate change, soil erosion, and deforestation are not new phenomena. They have a history connected directly to European imperial administration in Africa and Asia and to debates over the fertility of resource-producing regions of empire and the best method of administering indigenous populations. Malthus was certainly not alone!

There are two important points for the argument in this chapter that can be drawn directly from Richard Grove's summary of the research into the importance of colonization in environmental change. First is the contention that over the long run colonial management of resources has had a considerable effect that has long been overlooked in discussions of history and the politics of colonization and decolonization:

> Colonial ecological interventions, especially in deforestation and subsequently in forest conservation, irrigation and soil "protection," exercised a far more profound influence over most people than the more conspicuous and dramatic aspects of colonial rule that have traditionally preoccupied historians. Over the period 1670 to 1950, very approximately, a pattern of ecological power relations emerged in which the expanding European states acquired a global reach over natural resources in terms of consumption and then too, in terms of political and ecological control.[13]

Second, beyond this recognition is the one that concerns about environmental degradation and climate change were directly related to this expansion of control but within a larger cultural context that included attempts to appropriate resources for private capital and state use; worries that climate changes, droughts, and political disruptions might result from deforestation in the absence of vigorous conservation measures; and the cultural construction of Edenic and paradisiacal images of tropical climes as part of the larger European orientalization of the world.[14] In particular, these latter concerns related directly to the construction of game reserves and the preservation of species exclusively for European hunting. "The overarching process, however, was characterized by a process of drawing lines and boundaries. These both articulated the new assertion of control and arrogated the ecological realm to the state."[15]

The practices of enclosure, state management, and appropriation of resources thus have a long history directly connected to fears of political disruption and the dangers presented to these arrangements by the resistance practices of the dispossessed.[16] But this history is rarely countenanced by the neo-Malthusian accounts of scarcity-driven masses of people threatening modern political order. The assumptions of scarcity in contemporary thinking frequently ignore these histories and emphasize recent numerical increases in population as the problem, rather than understanding contemporary political problems as in part the consequence of these long-term geopolitical practices of enclosure and external control. The colonial economy is then taken as the benchmark against which management and development must occur, rather than as a series of complex historical structures constructed by previous developments.[17]

The links between climate and history are once again a matter of popular debate and scholarly analysis. Clearly there are relationships between environmental changes and human histories on the largest of scales, for example, El Niño cycles.[18] Nonetheless, of particular importance for the argument in this book is the focus on human actions in modifying environments, actions that are now at a scale that clearly are having global repercussions and may well be changing such climate patterns as El Niño. The human dimensions of global change are, at least in theory, within the realm of politics, a matter for collective human discussion and action. Hence clarifying the assumptions in the examinations of causal mechanisms and conceptual structures that examine these phenomena is an important scholarly task, but one where the history of science is anything but innocent.

GEOPOLITICS OR ECOPOLITICS?

More than most political scientists, Hayward Alker and Peter Haas understood the need to think about nature in the long term in their most useful account of the politics of the biosphere as an object of knowledge.[19] They argue that reading Vernadsky's ideas of a single biosphere and Braudel's historical formulations of the macropatterns of civilizations suggests more appropriate schemes for understanding the dimensions of global change.[20] Alker and Haas also suggest that the political framework that is invoked to think about global environmental change and the politics that are connected to this theme can be understood in terms of either geopolitics or ecopolitics. Geopolitics is derived, they argue, in part from Darwinian notions of biological competition and Malthusian-derived notions of the survival of the fittest. Applied to European rivalries and competition for territory through organicist models of states, this perspective found its apotheosis in the Nazi appropriation of Friedrich Ratzel's ideas of lebensraum to justify political aggression.[21] In contrast, ecopolitics is derived loosely from formulations of ecology as the study of the total relations of an organism with its surroundings, and from Vernadsky's understanding of the global biosphere as an ecosystem at the largest scale.

This recognition of the earth as a single dynamic system has subsequently been absorbed into concerns with sustainable development, articulated most obviously in the deliberations of the World Commission on Environment and Development.[22] It also appears in environmentalist thinking influenced by various schools of philosophical holism and in the recent discussions of the Gaia hypothesis.[23] Alker and Haas suggest that the geopolitical vision of nature in terms of competition between states at the largest scale won out during the cold war in the formulation of the containment of the Soviet Union as the most important political task for the West. This is related to the neglect of considerations of global ecopolitics. But the spatial imaginary of cold war geopolitics, and the perceived necessity to expand control over territory in the struggle with the other bloc, is of course not only a neglect of ecopolitics, it is also crucial to the processes justifying military preparation and industrial economic growth that have been an important cause of degradation of the natural environment.

While this focus on the macro scale may not generate the immediate policy-relevant scholarship that the Marc Levys of this world desire, it once again suggests that the questions posed about the politics of the global environment cannot be answered adequately by the neorealist assumptions that state-territorial actors are the key dimension of global politics and the source of the provision of adequate policy responses.[24] The crucial point is that concern about the global environment requires thinking about matters that transcend both the state and the conceptual tools of contemporary neoliberal scholarship. Notwithstanding some of its more flexible assumptions about the possibilities of social learning and regime construction, neoliberalism still operates on the state-territorial geopolitical premise that is so frequently part of the problem for those who wish to think about global problems in the biosphere.

INDUSTRIAL IMPERIALISM

Thinking in terms of changes to the biosphere, rather than geopolitical competition, suggests that the timescale for consideration of global change should not be taken for granted any more than should the geopolitical entities. In policy-oriented scholarly work, concerns about climate change are usually focused on no longer than a few decades. This is understandable in organizations like the Intergovernmental Panel on Climate Change, which has a specific rationale in providing advice and suggestions to international organizations and governments.[25] But such a focus may not be very useful in understanding the most important long-term trends. Ironically, because trends stress linear change, this is also probably the case in circumstances where rapid climate change may come unexpectedly.[26]

Anthropogenic global atmospheric change is driven most obviously by the widespread combustion of fossil fuels used in the processes that Lewis Mumford once so acutely called "carboniferous capitalism."[27] In biospheric terms, users of fossil fuels are transferring carbon from the rocks of the earth's crust into the atmosphere, reversing a long-term natural process of sequestering carbon from the atmosphere. In the process fossil fuel consumption is apparently disrupting the temporary stability of the global climate. That, of course, is not quite what the early pioneers of the steam engine, the British factory owners in Coalbrookdale, or the designers of the first iron bridge understood themselves to be doing. But understanding matters in terms of at least the last two centuries is necessary to deal with human-induced atmospheric change and its climatic consequences.

Thinking back two centuries to consider the implications of environmental change takes us back to the world of Napoleon, before the Congress of Vienna. The United States was but a collection of former colonies that had yet to forge a unitary identity. It takes us back to a period in history where the wealth per capita around the world was still loosely equitable prior to the dramatic rise in European power and affluence. The view from the end of the eighteenth century focuses attention on the dramatic changes that were to be wrought by industrial capitalism in combination with the forces of the second round of European colonial expansion. It also suggests that while European dominance was an important dimension of the geopolitical arrangements on the planet, it had not yet directly shaped large parts of the planet by its colonizing practices. India still had a textile industry, although the growing power of the coal-powered cotton industry in Britain, combined with imperial trade arrangements, was soon to put an end to it.

The geopolitical context of the last two centuries has changed dramatically while the industrial driving force of anthropogenic atmospheric change has developed into its current form.[28] The framework of contemporary politics is very different from that in the days of Napoleon, but the political processes understood in the conventional neoliberal and environmental security frameworks are largely limited to those that specify politics in modern statist terms. This is not to suggest the superiority of global civil society as an overarching framework to understand global politics, nor that there is necessarily any one other appropriate structure for dealing with the difficulties of climate change. Statist analyses, and a cartographic imagination of the nation-state and policies of regime construction, are unduly constrained in providing political and conceptual tools for grappling with contemporary ecopolitical difficulties, not least because modern states have long acted as development agencies and actively sought to promote carboniferous capitalism and its current form of car culture.[29]

Putting matters in historical context is necessary. Going back even half a century shows that the nascent political system of the United Nations of 1948, within which these matters could have been investigated then, is very different than in 1998, when the first draft of this chapter was written. Thinking about what we need to do now to anticipate ecological conditions in 2048 or 2098, never mind 2198, makes the point about the appropriate timescale for consideration. Independent states didn't exist in large parts of the world half a century ago, and so the institutional context within which such things as the United Nations Framework Convention on Climate Change are discussed did not exist either.[30] The problems of global change and appropriate institutional responses are more fundamental than is often recognized in international relations scholarship and related policy discussions.

PREINDUSTRIAL IMPERIALISM

Global change, ecopolitical structures, and their relevance to thinking about environmental security can perhaps be better understood if the timescale is extended back still further. Jared Diamond's magisterial overview of human ecological history is especially instructive in understanding the processes that have brought us to the current concerns with global change.[31] His suggestions lead to an understanding of the human expansion over the planet as a profound ecological influence long before industrial and colonial processes accelerated changes in the last few centuries.

From the destruction of numerous large animal species by overhunting, through the development of crop planting and the clearance of natural vegetation, to the emergence of domesticated animals, specific groups of humans have long acted in ways that either deliberately or accidentally changed both their environments and their relations to other neighboring humans. The evolution of some important infectious human diseases (smallpox, among others) as a result of animal domestication in particular places in Eurasia had implications for the populations who subsequently gained partial immunity to these diseases. This gave them crucial competitive advantages over nonimmune populations when they came into contact, as in the colonization of the Americas, where indigenous populations were drastically reduced by disease that spread well in advance of European settlement.

One of the most interesting arguments that comes from Diamond's reconstruction of ecological history is the recognition that the question of organized violence and the politics of destruction of weaker human societies is an ecological and geographical factor on the largest scale. Agricultural peoples have gradually, or sometimes not so gradually, displaced hunter-gatherer peoples. This process has accelerated enormously in the last half millennium where the expansion of Europeans into the Americas devastated their populations, removed much of the forest cover and indigenous vegetation, and spread a new agricultural ecology across much of both continents.[32] Critics of the colonizing practices of Europeans and of many contemporary states have long bemoaned the destruction of native cultures, peoples, and environments. The industrial imperialism of the last two centuries has merely accelerated and extended the scope of these changes.

The indigenous peoples destroyed in the process have often been nearly invisible to the dominant triumphalist narratives of modernization in the North. The expansion of European empires came at the cost of the destruction and obliteration of numerous indigenous societies, leaving, in Eric Wolf's still very apt phrase, "the people without history."[33] It is important to understand that this was often a matter of violent conquest with the prizes being land and other resources. As chapter 5 suggests briefly, the lost histories of indigenous suffering are now reemerging in the discussion of global environmental politics and, in the process, challenging how we think about both politics and environment.

Viewed in this long-term pattern of seeking land and food by conquest, the question of environmental security and the relations of violence to environmental change take on a very different significance. The encroachment of new ways of life into areas previously organized on different principles is usually a violent process. The relationship between violence and environmental change is thus not a new process. While many of the environmental security authors recognize some aspects of this lack of novelty, understanding both violence and deforestation in such places as Brazil in terms of the expansion of modernity suggests that the traditional interpretations of land wars on the agricultural frontier may explain much more about contemporary patterns of environmental change and violence than Malthusian assumptions about environmental change as a generator of conflict.[34] Seen as part of the processes of the expansion of modernity, political violence can be understood to be an intrinsic part of environmental change.

The erasure of indigenous populations is related to amnesia in the cartographic specifications of contemporary administrative practices.

As Michael Shapiro argues, the North American landscape is a conquerors' construction where the history of resistance is obliterated from maps and memories.[35] Whether as a direct colonial administration or through the commercialization and privatization of collective resource use, the practices of appropriating environments continue apace.[36] If these practices are understood as the driving forces of migration as well as of environmental and political change, then one gets a view of the processes involved in the environmental security problematique that is very different from conventional assumptions that environmental change may lead to insecurity. Viewed in historical and geographical terms, the practices of violence that are of concern in the environmental security discourse are the continuation of a pattern of conquest and dispossession that has a long and bloody history, recently accelerated by the processes discussed in chapter 3 in Karl Polanyi's terms as the "great transformation."

GEOPOLITICAL IMAGINARIES

These processes are now driven by the power of industrial technology that, both directly through fossil fuel emissions and indirectly through the dramatic expansion of rural exploitation to provide raw materials, has accelerated the processes of change. Thinking of nature as separate from the administrative practices of states and the cartographies of divisions and borders misses a crucial dimension of this dynamic. As Neil Smith has suggested, these are better understood as two facets of the same process, where nature and space are simultaneously produced.[37] A forest cleared by a new "settler" to plant crops is at once a change of land use, environmental change, and a spatial practice of enclosure supported by colonial land management arrangements. Space and nature are "produced" simultaneously. Considered in environmental terms, the division of ecosystems into separate parcels of land is a process of land use change that has dramatic environmental repercussions.[38]

Again the historical construction of these geographical matters is crucial. The point that many Southern peoples make, in arguing about the responsibility of those in the North for global warming, is that the historical record of Northern consumption is a crucial factor in the debate. Dramatic environmental changes have already occurred in the temperate parts of the world that Europeans colonized. Assuming that current environmental change elsewhere is a problem but that the historic changes wrought by colonization in North America are not to be considered obscures the relevant context for discussing global environmental security. Put in this geopolitical framework, the question "who is securing what where?" becomes especially pressing. Space cannot be taken for granted any more than environment can be.[39]

The conventional portrayal of geopolitical dangers conceals many of the important processes in the contemporary global political economy, in particular the fact that environmental degradation is often geographically distanced from the elites who benefit from the commodities produced in the process of degrading environments.[40] As the case of the environmental destruction of Ogoniland in Nigeria by oil exploration and extraction processes has recently made clear, the destruction of environments related to oil wells and other production activities often occurs far from the places where the petroleum is finally consumed.[41] Other mining operations and the disposal of at least some of Northern toxic wastes in states in the South show that the geography of pollution and degradation is much more complex than suggested in the simple models of zones of chaos and zones of peace.

Diamond's analysis also clearly shows that populations are in motion over the long term. Assumptions of spatially stable populations make as little sense as an ontological premise for politics as they do for understanding ecological realities on a scale necessary to grapple with climate change. Over the long course of human history, populations have moved and displaced other humans and their ecological niches; there is no reason to suspect that the twenty-first century should somehow be different even if the boundaries between states have now become very stable. Thus, a larger geopolitical vision sensitive to global change as an ongoing historical process, rather than as something new, suggests that the recent administrative boundaries of the postcolonial world are not useful as the framework for an analysis of either current problems or the long-term prospects for humanity.

INSECURITY AND CHANGE

Environmental discourse, as well as political discussion, is often caught in assumptions of stability or at least homeostatic equilibrium.[42] The important point for the contemporary literature about both development and environment is the recognition of the processes of change as the context within which these discussions must take place. The most basic concerns of the discourse on environmental change often betray their premises by posing change as a problem. This in turn assumes stability and the political status quo as the acceptable baseline for discussion. The contradiction is enormous, not least because development is so widely understood as providing much of the solution to current difficulties. Completing this circle is usually a matter of talking about political stability as the crucial dimension to be assured so that development can occur. The environmental conservation assumption of stability and preservation is frequently mapped onto political discussions of order and the lack of violence to formulate environmental security. But the dominant structures of the global economy, in its (violent) processes of expansion and disruption, are premised on change. The biosphere itself is also changing, and ecosystems are at best only temporarily homeostatic.

At the largest scale, the assumptions about biological conservation make sense in that changing the composition of the atmosphere may induce unpredictable rapid changes in the climate. Likewise, radically reducing the overall biodiversity of planet Earth, and in the process reversing the very long-term trends of evolution toward greater diversification, is also a dangerous gamble given the adaptable flexibility inherent in diversity. But these are not necessarily a useful set of criteria to apply directly to the smaller-scale considerations where disruptions and migrations are already in motion. Beyond these difficulties, formulations of environmental security often fall back on arguments that suggest that states are the only possible innovative mechanisms and providers of security, so they must be the most likely source of doing something useful about limiting environmental damage. This leads back to Matthias Finger's argument that the environmental security problem is driven substantially by states' attempts to render themselves secure in an industrialized world of security dilemmas.[43] They do so by appropriating natural resources to facilitate the industrial production of military systems.

This discussion is partly caught in the irony of increasing territorial stability and excludable definitions of property on one hand, and the dynamism of capitalism operating on a global scale on the other. Modernity is about motion and change, but it attempts to regulate itself in terms of fixed territorial boundaries and stable property lines.

Cartesian reasoning gives precedence to the stable, unchanging, and hence predictable. But from chaos theory to poststructuralism, contemporary thinking suggests ecological understandings of change and flux as more appropriate ontological categories precisely, and ironically, when territorial boundaries are apparently becoming more fixed than before.[44] Environmentalists have frequently argued that there is a fundamental ontological mismatch between states, their practices of sovereignty, and the ecological crisis.[45] But the crucial points in this chapter, about the historical dimensions of environmental change as a practice of violence and the geographical understanding of the historical mobility of people as anathema to the assumed political containers of the contemporary international system, suggest that the current debate over environment and conflict is caught in a series of disciplinary and institutional limitations that fail to grapple with either the scale of the problem or its historical trajectory.

The difficulties of understanding both politics and ecology as processes in motion, rather than as stable entities in need of securing in the face of change, are considerable. But as later chapters will make clear, any substantial engagement with the discussions of the environmental security problematique has to take these matters seriously. Specifically, the dangers of environmental determinism and neo-Malthusian assumptions about production limits in particular environments need to be countered by a more sophisticated political ecology that understands environmental change as a series of complex social processes in specific geographical contexts. In the process, simplistic assumptions about the efficacy of states, markets, civil societies, or projects of ecological modernization to provide solutions to environmental difficulties are put in doubt.[46]

There is absolutely no good reason to suppose that the poor majority of the world's rural population do not have the skills and capabilities to manage local resources and learn from the ecological conditions that they live within.[47] Getting away from neo-Malthusian assumptions about aggregate environmental resources to focus on the specific forms of ecological use and reconstruction used by rural peoples suggests that in many circumstances population increases in agricultural areas can lead to an increase in forest cover and the careful planting of trees to facilitate sustainable local production.[48] To naturalize the fate of the poor is to perpetuate many of the colonial practices of the past and to leave unchallenged the historical amnesias that occlude the political patterns of property appropriation and injustice that are part of the cause of so much contemporary misery.

Institutional frameworks, local customs, property relations, and local agricultural innovations are all important, and neo-Malthusian assumptions of rapid deforestation based on generalized assessments of environmental resources are frequently part of the problem. Likewise, there are compelling arguments against a politics of blaming the fecundity of poor women and women of color as another part of the problem in global environmental concerns.[49] However, the dangers of falling back into a geopolitical discourse of local virtue and distant perfidity has to be resisted in such generalizations—traditional methods are not necessarily benign. The point here is that the assumptions of disconnection are no longer tenable. We cannot deny the interconnectedness of the fates of people in different places in the biosphere.

Environmental change simply may not be the most important variable with which to start an analysis of environmental security. If environment is understood as an ongoing cultural process, security becomes much more difficult to invoke as an overarching policy objective. Focusing on migration and ecological disruption as the human condition, the dispossession of less powerful peoples as a long-term process accelerated and extended in the history of European expansion, carboniferous industrialization, and contemporary globalization offers a very different history and a more comprehensive causal sequence for understanding environmental insecurity.

As a result, in addition to reaffirming the importance of thinking critically about the spatial limitations of conventional international relations writing, the analysis in this chapter suggests four specific responses to contemporary formulations of environmental security. First, questions of environmental security are new only in the sense that the post–cold war political situation allows them to be discussed in a manner that the earlier geopolitical rivalry precluded. Ecopolitical matters are on the agenda, but conflict and environmental change have been intimately connected in the expansion of modernity for centuries. Second, the environmentally benign democratic peace, however desirable it may be as a policy goal, is premised on a (geo)political ecology that needs to be explicitly worked into the model. Questions of resource production, use, and trade and the specific geographies of environmental degradation complicate the state territorial framework of such theorizing. Incorporating these matters into the analysis also suggests, third, that the links between inter- and intrastate conflict in the analysis of environmental security need further elaboration in ways that bring environmental change directly into such models. Fourth, the whole question of the geopolitical premises in such analyses cannot be ignored if the theoretical analysis is to be robust enough to include both matters of environmental change and the politics of security in ways that grapple with the sources of environmental change at appropriate scales and in adequate complexity.

Ecopolitical considerations require that ecology and environmental history be taken seriously. While decisions about humanity's future are clearly political questions, the contextualizations in which they are thought about, debated, and decided need much more careful attention than has so far been the case in most discussions of environmental security.

5

Imperial Legacies, Indigenous Lives

Whether the colonist needs land as a site for the sake of the wealth buried in it, or whether he merely wishes to constrain the native to produce a surplus of food and new materials, is irrelevant; nor does it make much difference whether the native works under the direct supervision of the colonist or only under some form of indirect compulsion, for in every and any case the social and cultural system of native life must first be shattered.

KARL POLANYI, THE GREAT TRANSFORMATION

However, there is little doubt that it was in the imposition of new forms of land-designation, as between private and public and in the interruption of customary methodologies of interaction with forest, pasture and soil that colonial states (and post-colonial states effectively modeled on them) have exercised the most intimate and often oppressive impact on the daily lives and ways of production of the rural majority throughout much of the (especially tropical) world. This has been implicitly borne out by the apparent frequency of episodes of resistance to this species of colonial impact that have taken place throughout the period of the expansion of the capitalist forms of economic and political control.

RICHARD GROVE, CLIMATE AND EMPIRE

THE REST AGAINST THE WEST

Paying careful attention to the many interconnections across the taken-for-granted spaces of the geopolitical imagination is necessary to understand "global" problems. The historical evolution of patterns of interconnection is likewise crucial. This much was established in chapter 4. But the historical connections and the patterns of dispossession have further implications for understanding both environment and security. This line of inquiry suggests that the "great transformation" is still in process as globalization and development.[1] It also suggests that because the common assumptions of endogenous causes of social disruption and conflict are inadequate, the legacy of colonial disruptions may explain how the processes that are investigated in the literature on environmental security came to be understood in those terms.[2] This chapter explores these themes by first focusing on another article from the *Atlantic Monthly,* one that in many ways is a rejoinder to Robert Kaplan's analysis and that clearly suggests the importance of the geographical assumptions in this debate.

Long-distance migration and the likely social consequences are one important theme in contemporary discussions of Northern security mentioned only in passing in Kaplan's analysis.[3] Matthew Connelly and Paul Kennedy's later article in the *Atlantic Monthly* looked specifically at migrations of impoverished humanity in motion as the global order changed at the end of the cold war.[4] The environmental theme is of less salience in their article, which focuses more explicitly on demographic matters. In the context of current fears about illegal migration in both Europe and the United States, they look to Malthusian speculations about global demography and raise the question of whether demographic politics has to be played out in a geopolitical conflict between "the Rest" and "the West." In particular they focus on "the key global political problem of the final years of the twentieth century: unbalanced wealth and resources, unbalanced demographic trends, and the relationship between the two."[5] In contrast to Kaplan, who is concerned with the spillover from the wild zones to the tame ones, but who never looks seriously at international migration as a mechanism for this "danger," Connelly and Kennedy examine this geopolitical factor directly.

Where Kaplan relies on his eyewitness journalistic accounts to set up his larger discussion, Connelly and Kennedy start with Jean Raspail's controversial early 1970s French novel *The Camp of the Saints,* focusing on its dramatic story of impoverished Indians hijacking ships and setting forth across the oceans for France. The designers of the *Atlantic Monthly* use a dramatic cover illustration, framed once again in the spatial terms of a tension between fear and aspiration, to emphasize the theme of the article. It shows a pale-skinned suburban householder equipped with a spatula and wearing an apron emblazoned with the motif "home sweet home." Accompanied by his dog, he is standing on a patio beside a barbecue on which he is cooking sausages. The suburban ideal is marred only by the many dark-skinned faces, some clad in various "ethnic" headgear, who are looking over the white picket fence surrounding his yard. The text superimposed on the fence summarizes the theme of the article: "Whether it's racist fantasy or realistic concern, it's a question that won't go away: As population and misery increase, will the wretched of the earth overwhelm the Western paradise?"[6]

The article argues that, while Raspail is probably guilty of a variety of racist sentiments, the themes in this disturbing novel are very germane to current discussions of foreign policy and the focus in the United States on immigration. In particular, the relative decline of the European races in terms of total numbers of world population suggests the inevitable triumph of the former colonized peoples who will in the next few decades, as European populations atrophy, reverse the geopolitical patterns of North and South:

> (W)e are heading into the twenty-first century in a world consisting for the most part of a relatively small number of rich, satiated, demographically stagnant societies and a large number of poverty stricken, resource depleted nations whose populations are doubling every twenty-five years or less. The demographic imbalances are exacerbated by grotesque disparities of wealth between rich and poor countries. Despite the easy references that are made to our common humanity, it is difficult to believe that Switzerland, with an annual average *per capita* income of about $35,000, and Mali, with an average *per capita* income of less than $300, are on the same planet—but Raspail's point is that *they are,* and that a combination of push and pull factors will entice desperate, ambitious Third World peasants to approach the portals of the First World in ever increasing numbers.[7]

While the neo-Malthusian framework is in the presentation of the argument in terms of massive dislocations and migrations from the poor to the rich world, this article's conclusions are notably very different from Kaplan's geopolitical pessimism. It notes the arguments by the technological optimists who argue, in response to Kaplan's despair, that global economic indicators show widespread signs of optimism, but that this optimism is not in any practical way linked to the fate of the poorest billions of the world's population.[8] They also point out that while production has been globalized, the mobility of labor has not. Geographical restrictions on the mobility of workers are in dramatic contrast to the ability of transnational corporations to switch production and investments around the globe.[9] Even if the "techno-liberal" optimists are correct and growth does occur, it seems likely that given population growth the absolute, if not relative, numbers of very poor will increase.

Drawing on the elaborated speculations in Kennedy's earlier book *Preparing for the Twenty-first Century,* the article offers much greater recognition of the interconnectedness of global problems and proffers suggestions for policy initiatives that tackle poverty and related economic and environmental issues.[10] The scenario of desperate, impoverished people attempting to move to the affluent world and the unpleasant policy implications of trying to resist such migrations by force are merely hinted at. But unlike Kaplan, with his unexamined assumptions of environmental degradation, the geopolitical version of the neo-Malthusian scenario is not judged to be inevitable. Instead they argue the case for a new North-South political deal in which global cooperation is seen as necessary by political leaders. They admit that transcending partisan and national perceptions of political possibilities and difficulties may not be easy, but it is clearly necessary to deal with global problems.

In Virginia Abernethy's *Atlantic Monthly* magazine rejoinder to Connelly and Kennedy, she argues that development assistance to poor states often renders their populations more fertile by raising hopes that the development projects ultimately fail to deliver, hence aggravating the problem of population numbers.[11] If the political consequences of population growth are disruptive to the Northern geopolitical order that is judged to be the only acceptable one, then neo-Malthusianism acts as a powerful intellectual weapon in formulating policies to repress and politically control reformist demands for greater equality or economic redistribution. It can do so on the grounds that such policies only aggravate adverse demographic trends. When coupled to Kaplan's assertions that population growth is related to environmental degradation, the argument is strengthened.

If the more alarmist versions of some of Kaplan's arguments gain long-term credence in Washington, or if the formulation of politics in terms of the Rest and the West becomes prominent, then the dangers of a new cold war against the poor are very considerable. Discussions through the 1990s of illegal immigration to the United States proposed that the solution is increased border guards, denial of services to immigrants incapable of proving legal residence, and deportations, which suggested that the geopolitical imagination of spatial exclusion was dominating the policy discourse once again.[12] This geopolitical imagination has been frequently coupled to assertions of cultural superiority and ideological rectitude in the form of various articulations of moral certainty. Through the course of the cold war, and subsequently in the 1991 Gulf War, these formulations have fueled arms races, the global politics of deterrence, and security understood in terms of violent containment and military superiority.[13]

But such policy measures may not be effective in dealing with the long-term patterns of global impoverishment and demographic change. Connelly and Kennedy note that the conclusion of Raspail's novel describes French troops deserting rather than using sophisticated weapons to massacre defenseless migrants or torpedo their decrepit ships. If military solutions are invoked and implemented, the specter of "global apartheid" with large-scale ethnocentrically justified geopolitical divisions will take another few steps forward.[14] As the history of the last few decades in Southern Africa suggests, such arrangements are not sustainable in the long run, either in terms of political institutions or in terms of the environmental consequences of very badly skewed resource allocations. In particular, it is clear that even the sophisticated geographical administration of apartheid, backed up by a substantial military capability, could not constrain the migration of poor people to the cities, which rendered the system less and less workable.

POLITICAL ECOLOGY: GLOBAL PROCESSES, LOCAL HAZARDS

The concern with environmental degradation is often imprecisely attributed to population pressures and to pollution that is understood as an inevitable part of development. Population is related to environment in numerous ways, but the relationships are mediated by complex social and economic arrangements that need detailed attention. While population increase is a factor of importance in many locations, it is not necessarily a cause of either environmental degradation or acute conflict in many places, including Rwanda, where simplistic generalizations incorrectly specify population increases as a major cause of environmental degradation and conflict.[15]

Political ecology investigates environments and social conflict in specific locations in relation to global economic processes. This approach suggests that simple assumptions about processes of environmental degradation are inadequate.[16] The research also reveals that complicated social processes determine the environment-specific and resource-related issues of access that underlie questions of environmental scarcity, and that gender in particular plays a crucial role at the immediate scale of lived experience.[17] Above all, the political ecology literature suggests that the physical and environmental conditions characterizing a given locality need to be incorporated in an analysis that links ecological factors through the specific institutional context of land ownership and marketing systems to explain the dynamics of conflict. Generalizations about environmental scarcity are always in danger of equating very different ownership patterns and the specific ecologies that provide subsistence. Likewise, environmental degradation comprises a complex interactive set of processes, many of which are reversible, some of which are permanent.

As shown in chapter 3, some of the environment and conflict analytical frameworks do adopt facets of such an approach in the search for causal factors when investigating the growing relative scarcity of environmental resources or the problems of capturing resources at the cost of ecologically marginalizing those thereby dispossessed. Understood in terms of the political ecology framework, conflicts about rivalries for direct control over resources are about matters only indirectly involving environmental degradation and, even then, not in ways that would make such degradation the driving cause of those conflicts. In political ecology, larger-scale global economic forces and commodity markets are integral to the processes that drive modernization and commercialization, both of which occasion displacements.[18] So interpreted, environment itself might become an intervening variable in the political processes that lead from resource capture to conflicts precipitated by the actions of the marginalized who are simply resisting dispossession or expropriation.[19]

If processes of resource capture are explained in their larger political economic context, within a discourse sensitive to changing resource ownership and control, then understandings of conflict, environmental change, and policy issues are likely to become much more insightful. But they do not fit easily into the framework of environmental degradation leading to conflict. The policy implications are also very different, not least because the modern assumption of scarcity as the general ontological condition of humanity is no long-er the unquestioned premise for the analysis.[20] Scarcity, or notions of environmental stress, are now understood as in part political pro-cesses, and as such they are amenable to social change in more com-plex ways than those premised on conventional resource manage-ment assumptions.

In analyzing these matters in terms of the relationship between agriculture and violence, Indra de Soysa and Nils Petter Gleditsch summarize the situation in many poor states where urban policy and power increase rural poverty and may lead to violence and political rebellion. They suggest that governments' primary preoccupation is maintaining power and that the manipulation of food prices to facili-tate control of the urban population is a major factor in the politics of underdevelopment. Keeping food prices low reduces incomes to the smallholder farmers. Tariffs and import restrictions inflate local currencies and benefit large crop exporting sectors while increasing incentives to produce cash crops rather than food crops. Small farm-ers and landless peasants are thus squeezed off the land while urban elites, state workers, and export agriculture benefit. Peasants moving to the cities in search of work reduce urban wages and support the urban bias of policy makers, further reducing the incentives for food production or land reform.

> Under such conditions, it is not surprising that historically the foot soldiers of rebellions against states have been landless peasants and their poor cousins recently moved to the urban slums. Rent seeking and urban bias also have implications for violence through the crea-tion of patrimonial politics, patronage and the destruction of social capital. Clientalism creates vertical ties of dependency between patron and clients at the expense of horizontal ties of association, which are the foundations of the effectiveness of government and of satisfaction with government performance.[21]

All this is also coupled to the complex workings of the inter-national financial system, the frequent encouragement of export agri-culture as a development strategy, rural distress, elite manipulation of international aid, and structural adjustment fund conditionalities. In sum, this suggests a complex situation of rural dispossession and political violence. The Rwanda genocide of 1994 has complex roots, but understanding that these factors contributed to the violence is important if the simplistic portrayal of these events in Malthusian terms is to be avoided.[22] The crucial point is that the specific factors in one state are related to the larger political economy within which food, land, and violence are enmeshed. To fully criticize the Malthusian assumption of scarcity requires understanding that the political ecology of poverty and violence is about more than an endogenous situation within the bounds of a single autonomous space, however important the specific actions of elites are in particular states.

A similar argument can be made by drawing from the literature on disasters and their management. While hazardous natural phenomena such as storms, floods, tornadoes, droughts, and volcanic eruptions may or may not be gaining in frequency or severity, their impacts have been swifter in coming and more damaging during the last few decades. Many people are living in more vulnerable locations, such as on lowlands or steep slopes or in the path of fires in Southern California. While human activities—from deforestation to dam building—do affect the parameters of floods and droughts and may indirectly increase vulnerabilities, many casualties are simply a product of physical insecurity engendered by poverty and dispossession that forces people to live in dangerous locations.[23] The processes of marginalization are often compounded by ethnic rivalries and state inaction. But the crucial point in the contemporary disasters literature is that both vulnerabilities and responses to disasters need to be understood as cumulative social processes.[24] The specific links between the components of the poverty that produce vulnerability are important considerations in the analysis of disasters and of the responses to them. Blanket assumptions about scarcity often fail to clarify this point.

The political dimensions of disasters are related in a number of ways to the more traditional considerations of national security.[25] Disasters can offer numerous opportunities for social change, perhaps directing emergency and reconstruction aid to the dominant groups of the damaged society and thus further marginalizing its poor. For example, in the designation of aid for housing reconstruction, renters and squatters often do not qualify for aid, whereas the wealthy holders of title to lands and buildings get additional advantages in the reconstruction process. The diversion of aid to elites has been a matter of political contention after many disasters in Central America. The collapse of the Somoza regime in Nicaragua in the 1970s is a clear example; outrage over the diversion of aid swelled the ranks of the re-gime's opponents. Similarly, the failure to deal with famine in the early 1970s by the Halie Selassie administration in Ethiopia undoubt-edly hastened its demise.[26]

GREED NOT GRIEVANCE?

The question of elite appropriations of resources and the use of vio-lence to retain power in disaster situations can be further stretched to suggest that the Malthusian assumption within the environment and conflict literature formulates things backward. De Soysa's sum-mary in a recent paper title captures the essence of the argument, pos-ing matters in terms of either a "shrinking pie" or a "honey pot."[27] The "shrinking pie" is an obvious reference to the scarcity-driven conflict model. The argument that elites fight over control of the reve-nues from a resource stream implies that violence occurs in resource-rich areas, or "honey pots," rather than in areas of scarcity, where the "pies" available for distribution are shrinking.

This connects to the discussion about ingenuity and adaptation because if there are resources that do not propel people to adapt to shortages or provide any incentives to innovate economically, then ingenuity will be absent. Living off available wealth is an easier eco-nomic strategy for elites who control the wealth than inducing the innovations that will move the economy into competitive manufac-turing. The assumption is that an economy is successful if it moves out of resource production and into manufacturing, a reprise of stan-dard assumptions in most development thinking. The argument that plenty is a curse on innovation and development is receiving more attention recently as the implications of contemporary patterns of violence are investigated beyond the constraints of the hypothesis that environmental degradation leads to conflict.

The themes of modernization and the penetration of development into remote regions in Gunther Baechler's work, discussed in chap-ter 3, raise the question of how industrialization in Europe might in-cite conflicts on the periphery of the world economy. Resource scar-city has been assumed to be the cause of Malthusian violence and forms the basic premise of the discussion in some of the contempo-rary literature making the links between conflict and environment, but de Soysa argues that violence and access to resources might be re-lated to abundance rather than scarcity. This line of analysis suggests that fights between elites to control the revenues garnered from the extraction and export of resources may better explain the violence, in Africa in particular, than Malthusian assumptions of resource shortage. Once again, putting the historical legacy of colonialism into the analysis emphasizes the pattern of resource extraction and violence that has occurred as indigenous arrangements were disrupted and populations forced into the commodity economy by colonial taxation arrangements.

De Soysa suggests that a high proportion of natural resources exports are related to civil war more than to scarcity.[28] The contention here is that it is easy to take control of a resource stream of primary goods and that the revenues that can be extracted from the stream make conflict a worthwhile gamble for political leaders. Coupled to the availability of many unemployed and uneducated young men, with little to lose in joining in greed-motivated rebellion, the possibilities of violence and civil war can be high. But the analysis needs to be more complex than this because it is possible to argue that resource-dependent states are facing scarcities that are the cause of conflict.

De Soysa uses measures of total per capita natural resource abundance, disaggregated into measures of renewables and nonrenewables instead of the share of primary resource exports, and concludes that abundance is positively correlated with conflict. He also concludes that human and physical capital rather than natural resource availability are predictors of economic growth. Resource scarcity, at least at nationally aggregate levels, is not correlated with conflict in this analysis. While the downward spiral of violence preventing ingenuity, which Homer-Dixon is concerned about, may operate in some places, de Soysa's suggestion is that ingenuity is stifled by abundance rather than violence. Greed rather than grievance is more usually the cause of conflict concerning resources. Many of the poorest states in the world are those most heavily dependant on primary product exports.[29]

Whatever the source of the violence in specific places in Africa, Malthusian subsistence crises usually do not provide a credible explanation.[30] There are land shortages in some places, but these are not the cause of much of the violence in Africa. Paradigmatic is the case of the colonization and destruction of what became the Congo by Belgian colonists, and King Leopold in particular, in the late nineteenth century. As Adam Hochschild details, the extraction of rubber and subsequently other commodities effectively enslaved a large part of the native population and cut them off from traditional patterns of subsistence.[31] The disastrous famines and destruction of the population eventually endangered the production of rubber and the whole colonial economy. The legacy of these extractive practices was continued after independence when President Mobuto built a personal fortune in Swiss banks and European real estate while the limited infrastructure of his country crumbled. Western support for his regime continued despite the obvious inequities and the persisting political violence. Not surprisingly, war finally brought an end to this regime, but the subsequent civil war there has been fed by the continued supplies of arms and armies from neighboring states.

The argument that political violence is directly related to the extraction of resources, rather than resource scarcity, is supported by analyses of the political ecology of conflict in Southeast Asia. "Local communities are caught in the conundrum of depending on natural resources while being largely marginalized from the political practices, often illegal and predatory, of governments and extractive industries that profoundly impact on the local resource bases."[32] Arguing that the conventional development assumption that wealth extracted from the natural resource base will flow into socioeconomic progress and benefit the national population does not hold, at least in the cases of Burma and Cambodia, Kirk Talbott and Melissa Brown point to the enrichment of local elites and the resulting corruption and violation of what laws exist to prevent environmental destruction. Military and political organizations frequently divert revenue streams to support their activities, simultaneously reducing the capabilities of state agencies and ensuring that forests are mined to support political and military struggles. In Burma this destruction is interconnected with the complex patterns of insurgency and the repression of indigenous peoples by the military rulers. Their analysis of these dynamics leads the authors to conclude:

> Logging is central to the downward spiral in the region; corruption among the political and economic elites leads to rogue logging, which in turn fuels a further disintegration of the structures of civil society and good governance. As such, the accelerated deforestation caused by excessive and illegal logging severely undermines the chances for the development of independent judiciary systems, political accountability, and effective law enforcement. Long-term, sustainable natural

resource management will not occur without these, the tenets of civil society.[33]

Colin Kahl develops similar arguments about elite violence in Kenya and the Philippines, showing that elites often use violence to maintain power and suggesting that violence in rural areas is frequently a matter of urban elites struggling to maintain control there.[34] What is clear in reading all these analyses is the need for careful anthropological field work sensitive to the details of historical cleavages and power structures in particular places; not all places with resources prone to easy looting or revenue stream control dissolve into conflict.[35]

INDIGENOUS DISPLACEMENTS

A further extension of this line of argument can be made in terms of the destruction of livelihoods and lands by development projects on indigenous people's lands, such as mining, logging, or damming, which links back to Baechler and the ENCOP analysis discussed in chapter 3. Postcolonial states have often continued to practice many of the once-maligned policies of their colonial predecessors whose paths to development they now try to emulate.[36] Peoples indigenous to the jungles of the tropical world are considered to be primitive, whether in Malaysia or Brazil. Modernity deems that their traditional ways of life are not worth taking seriously.[37] Their tribal cultures and experiences are dismissed as unscientific and hence not useful beyond providing pharmaceutical companies with the sources of traditional remedies that can be "developed" to provide drugs for the medical establishments of the industrial powers of the world.

The defining language here is, not surprisingly, politically loaded. The terms "jungle" or "rainforest" connote biologically similar but politically different entities. Not many people worry about the fate of "jungles," but once "rainforests" are turned into a cause by the environmentalists of the North, "the Amazons" become a treasure to be "protected" against "the state" claiming sovereignty over it.[38] The focus by anthropologists on the human rights of "local" peoples and their struggles against encroachment adequately clarifies the question of what exactly is being rendered secure.[39] In most cases, what is to be secured is neither intact indigenous societies nor benevolent settings that support their ways of life. These are not in need of securing, according to modern theories of development; they are in need of modernization, education, and all the panoply of modern modes of being. While traditional cultures may be of use to the tourist trade, "primitive" cultures are the object of numerous development projects, and their traditional usufructuary practices are then quickly displaced by the enclosures and resource appropriations of modernity.[40]

Modernization is by definition a disruptive force of change, so the spread of modernity not surprisingly sometimes breeds conflict. Violence on the frontier is nothing new in the history of colonization.[41] It is a very old pattern that has little to do with Malthusian interpretations of poverty driven by overpopulation. In some cases resistance to the exploitation of indigenous resources has the classic makings of the independence struggles that replay the history of decolonization conflicts.[42] In an interesting inversion of the conventional assumptions about state security and territorial sovereignty, Bernard Nietschmann takes this theme to its logical conclusion and argues that through much of this century the industrialized world has effectively been involved in a war of destruction with the indigenous cultures of what is sometimes called the Fourth World.[43]

The discussion of environmental politics in terms of states, regimes, and the territorial spaces of North and South is obviously much more complicated than conventional cartographic imaginaries can adequately accommodate. But even if the inadequacies of such categories are built into discussions of international politics, if the North-South discussion and the history of ecological change is taken seriously, there is much more to the matter of environments and their histories that is forgotten. Perhaps most obvious is the point that the neo-Europes, to use Alfred Crosby's term for the temperate climate regions colonized by Europeans, were previously inhabited by indigenous peoples, often for long periods before the arrival of colonists.[44]

Some of these peoples were completely eradicated by direct violence, disease, and the destruction of their lands and sources of livelihood. But many were pushed onto marginal lands or displaced to become the underclass in growing urban centers. Their voices are now being heard in discussions of environmental politics. Their presence further complicates the political geography of ecology, raising questions about what the geography of sustainability or environmental security might look like, and ensuring that blanket statements about national environmental goals, whether couched in terms of development or security, are challenged by detailed specifications of where the environmental disruptions will occur. Even more than this, native activists powerfully challenge the assumptions of environment as a politically innocent category.

DUMP BEARS AND INDIAN AGENTS

Native peoples in North America and elsewhere challenge the assumptions of national security by pointing out that national security for the state is frequently maintained by rendering native peoples very insecure.[45] National zones of sacrifice for uranium mining, weapons testing, and military training frequently turn out to be on aboriginal lands. This is not coincidental; the remaining lands designated for native peoples are frequently remote from large urban centers and often under direct control by central governments. Whether in fights over access to resources and land or in attempts to minimize the environmental disruptions to traditional hunting territories, native peoples are often in the way of both the military plans of large states and the aspirations of resource companies.[46]

Indigenous peoples, militarily weaker than the invading colonists and vulnerable to diseases brought by settlers, have historically been the victims of many forms of colonization. In the process of their conquest, their histories have nearly been forgotten. Nonetheless, the lessons about resource control and security implicit in these histories may have much to teach about the contemporary conceptualizations precisely by recovering what has been lost in the celebrations of conquest and assumptions of assimilation. The aboriginal histories also raise fundamental questions about modern identity and its assumptions of how to secure that identity, which will be discussed in more detail later.

One empirical example drawn from Canada illustrates the most salient themes for the argument. The Mi'kmaq people's traditional territory stretches across a large part of what subsequently became known as the maritime provinces of Canada: Prince Edward Island, Nova Scotia, and New Brunswick. Although many of the Mi'kmaq were deprived of access to their traditional lands, resources, and hunting grounds and confined to a number of reserves in the twentieth century, in common with so many other native populations in North America, some of their culture has survived the assimilationist attempts of the Canadian state and the Catholic Church.[47] The northernmost part of Nova Scotia consists of the island known in English as Cape Breton. In the Mi'kmaq language the island is known as Unama'ki, a place where one of the key figures in the Mi'kmaq cosmology, Kluscap, rests in a cave until he will awake to come to the aid of his people in their hour of greatest peril. In the late 1980s a local mining company, Kelly Rock Ltd., planned a major quarry on top of the mountain above the cave. Local Mi'kmaq activists as well as non-Mi'kmaq environmentalists opposed the proposal.[48]

In doing so the Mi'kmaq joined with many other native peoples across North America in tackling their marginalized plight through struggles that can be broadly designated in terms of the environment.[49] And yet, as noted in the preface to this book, the terminology doesn't quite fit the activists' self-understandings of their situation. These native voices from the margins challenge the designations of environment and do so in a way that directly confronts the colonial legacy of the past and suggests that the same patterns of resource expropriation continue to fundamentally shape the fate of native peoples in many places. These geographies are clearly understood in Unama'ki where one Mi'kmaq activist, interviewed in 1997, drew on a powerful image to link environmental destruction with the disruptive consequences for both nature and people. Referring to the destruction of the forests in British Columbia (B.C.) and the marginalization of indigenous First Nations peoples in the slums of Canadian cities, the following statement captures the key geographical processes in especially evocative prose:

I had the good fortune to see Meares Island in B.C. where the giant trees are. You see the whales off the coast. Beautiful.

But I also had the misfortune of going into the clear-cut areas and seeing the giant stumps. Giant, giant grandfather trees that stood there for years and years, now just being logged out and cut indiscriminately. Whole mountainsides just bald. Shaven bare.

I've seen the bears in the dump, and they're labeled nuisance. But the bears are not in the dump because they want to be or that they like the dump, the garbage. They are there because their natural home is no longer there. There's no more berries, no more salmon, no more plants. And as a result they are driven down into the cities, the towns, the garbages.

Any being, whether it be the four-legged variety or two-legged variety, would rather eat fresh salmon and berries than rotted garbage. Any being. But one day they are driven down into the dumps and they are labeled nuisance, just like my people are labeled nuisance when you see them in Ottawa, Toronto, Vancouver, Winnipeg.

When you meet a First Nation on the street, it's just like the bear. He's driven down to that dump because there is no more trapline, there's no more berries, no more salmon, there's no more forest. And as a result they are driven down to the big city dumps and they are looked upon as nuisance, just like the bear.

And if we're to try to put a balance back into nature, then we have to stop the destruction. We have to get the salmon back into the rivers, the berries back, the trees. So that the bears and the indigenous people can continue the way the Creator meant us to.[50]

There is a simple but powerful geography to these matters, where the resources to maintain the consumption patterns in cities depend on extraction at a distance and where the growing impact of urban centers disrupts both people and animals. The analogy between dump bears and native peoples, both of whom are literally displaced, emphasizes the transformative impact of the extraction of resources for distant consumption. But it also points to the disruptions of ecosystems as part of the processes of modernization. Economic growth is premised on the colonization of distant environments. The Kelly Rock company plan to dig a quarry on top of Kluscap's mountain in Unama'ki, to supply aggregate for road building and other construction projects on the East Coast of North America, merely continues the long-term pattern of resource extraction that has been the history of this island since the early arrival of European fishing vessels five hundred years ago, long before British colonists and the subsequent Canadian government established control over the native peoples and instituted a system of Indian agents to administer resource extraction and control native populations.

But the question of how to respond to this legacy raises a further crucial point in considering how to think about these matters in terms of environment or security. The often uneasy alliances between environmentalists and native peoples, as was the case in the (ultimately successful) opposition to the Kelly Rock quarry proposal, pose questions of who is protecting what in "environmental" campaigns. Native activists, struggling to adapt and cope within the marginal circumstances in which colonization has placed them, often focus on native peoples' issues as their first priority. In activist academic Ward Churchill's terms, this is an indigenist politics, one that "draws upon the traditions—the bodies of knowledge and corresponding codes of value—evolved over many thousands of years by native peoples the world over."[51] The colonial assumptions in environmentalism are rarely more clearly expressed than in the following distinction drawn by one of the activists in Unama'ki:

> I never claim to be an environmentalist. I claim to be a First Nationist, which differs from an environmentalist. Environmentalists . . .—they're damn good people, don't get me wrong—but I find they are the ones that appoint themselves guardians over the forests, guardians over the animals, the fish, the water. They're the guardians, the superintendents, the Indian agents of all this.
>
> The First Nationists harvest the rivers, the forests. We harvest. We are still part of the ecosystem. We will harvest the moose, the salmon, the trees, for what we need for that day to survive. We won't harvest a mountain of trees so that we can make our home. We will harvest what we need. And on the same token the tree will harvest us if we're not careful. The bear will harvest the First Nation person if he is not careful. The wolf will harvest us, the ocean will harvest us if we are not respectful. The rivers will harvest us.
>
> We are not above them and we are not beneath them. We are still part of this ecosystem. And that's where it differs between environmentalist and First Nationist.[52]

The environmentalist arguments frequently focus on questions of the appropriate use of resources and the preservation of nature, species, and habitats. This native activist understands that the construction of "environment" is a colonial understanding, one that operates on the urban assumptions of an external nature whose resources are to be managed, rather than a context, place, or home that is to be lived in. But the insight is especially telling in light of Richard Grove's extensive historical investigations into the origins of contemporary environmentalism, which, as noted in chapter 4, are linked directly to matters of colonial administration and anxieties about climate, deforestation, and much else.[53]

The cultural construction of nature as external is of course an extension of the etiology of the term "environ," which literally refers to that which surrounds, and historically to that which surrounds a town.[54] As the scale of the global economy expands, and as the population in cities makes ever larger demands on distant rural resources, the question of the appropriate designation of these processes becomes ever more critical. In Arturo Escobar's terms this is so because "As they are incorporated into the world capitalist economy, even the most remote communities of the Third World are torn from their local context, redefined as 'resources' to be planned for, managed."[55]

In discussing environmental security, the expansion of urban expropriation of rural resources has to be worked into the analysis if the appropriate geographical understandings are to be made part of the discussion. As chapter 4 has made clear, getting the geography of environmental security wrong does not help clarify matters. Geopolitical reasoning may be a powerful mode of raising political concern about security issues, but as a mode of thinking intelligently about contemporary social and environmental processes, it leaves much to be desired, precisely because it so frequently perpetuates the patterns of development thinking and the geopolitical assumptions of separate competing polities that are the cause of so much difficulty in the first place.

But before extending the argument about dump bears, Indian agents, and displacements in chapter 7, it is necessary to flesh out the argument in this chapter about the importance of contexts and connections by explicitly rethinking the geographical premises of environmental security. If neither territorially oriented views of sovereignty nor the conventional understandings of development are adequate for understanding the flows and interconnections of the global economy, as well as the biosphere, then other modes of thinking are necessary in specifying political relationships in ways that get at the social processes that matter in causing conflict and environmental degradation. Beyond that, theoretical ideas drawn from the science of ecology, rather than the literature on environmentalism, may also offer useful ways to reformulate security. Chapter 6 examines alternative geographical specifications, and chapter 7 considers the implications of ecology and the possibilities of integrating ecological thinking into the rethought geographies of insecurity.

6

Shadows, Footprints, and Environmental Space

If one is to think seriously about the world one must have recourse to a spatial vocabulary. In the absence of better alternatives, exhausted meta-geographical concepts will creep back into use, even in acutely critical post-modern texts.

MARTIN LEWIS AND KAREN WIGEN, *THE MYTH OF CONTINENTS*

P...ther, what is needed is a geographical imagination that takes places seriously as the settings for human life and tries to understand world politics in terms of its impacts on the material welfare and identities of people in different places.

JOHN AGNEW, *GEOPOLITICS*

COLONIAL CONNECTIONS

The geographical patterns of the colonial experience in the Americas were about the conquest and dispossession of indigenous peoples in part, but also the extraction of resources to support growing European economies. Silver and gold from South America were important in the early expansion of Spanish power in Europe, which gave way to Dutch, French, and British dominance. Many of the wars between these powers led to changes in the ownership of colonies, and the history of Cape Breton is no exception. Struggled over for a century by the British and the French empires, the Mi'kmaq warriors were active participants in the military campaigns. The tourist trails in Nova Scotia are about scenic coastal vistas and sports fishing, but they are also a series of sites commemorating this history with reconstructions of old forts and interpretative plaques about colonial facilities.

Historically, the fishing on the Grand Banks off the shores of Cape Breton made Port Louisburg especially important. As one of the largest ports on the Atlantic coast in the early eighteenth century, it was seen as a trading rival by the growing colonies to the south. Boston merchants were enthusiastic supporters of its destruction in the 1740s and key organizers of the military campaign that subsequently took the port by siege. The interconnections between access to North American resources and the European wars are direct, with the legacy of French power still a prominent part of Canadian politics. The very existence of Canada and the United States has to be understood in these terms if the political geography of North America is to be intelligible.

Just as most of the political spaces of the Americas gained some formal independence in the late eighteenth and early nineteenth centuries, the growing European economies expanded their military power and economic impact in other parts of the world in the second phase of European imperial expansion. Again the economic importance of various resources was crucial, and plantation agriculture spread to Asia to provide rubber, coffee, tea, spices, cotton, and all manner of other commodities. After independence such trading patterns were usually maintained. As chapter 5 has suggested, many postcolonial societies are still tied into patterns of export that persist from colonial times. Forests and other resources are stripped in the South to make commodities available in the North.

However, in terms of the potential for violence in the disputes over environmental resources in the South, international resource flows may be only indirectly involved in many cases.[1] Part of the reason for the lack of connection is the crucial finding in the environmental conflict literature that North-South wars over resources are not likely. Environmental conflicts are diffuse and subnational. Even when resources such as minerals and oil were ostensibly a matter of international relations in the cold war period, research showed that access to resources was nearly always part of a larger economic and ideological conflict rather than a driving force in international relations.[2]

But there are enough center-periphery conflicts in the case studies on environmental conflict to make it clear that the indirect effects of resource flows are a crucial part of the larger picture. Violence over the political and environmental disruptions caused by the extractions of oil in Ogoniland, copper in Bougainville, or aggregate from Kluscap's mountain in Nova Scotia is unavoidably part of the larger processes that environmental security research has to investigate.[3] Where resource flows do appear in the international relations literature, their significance for the large-scale environmental changes that are crucial to the politics of global environmental change is often obscured because of the focus on regulations and state administrative procedures in regime construction. While the necessity of administering resources is understood, the direct connections to ecopolitical changes within states are often much less than clear.[4]

Using state-territorial entities as the basis for analysis of the causes of environmental destruction focuses attention within the containers and in doing so offers a very selective geographical understanding of the processes of global change, resource use, and degradation. The geography of population displacement and environmental degradation is much more complex than formulations of politics operating within the limitations of the territorial trap usually allow.[5] In his suggestions for improving the conceptual framework used to examine the relationships between environment and war, Gleditsch points to the necessity of connecting inter- and intrastate violence.[6] But the connection of nonviolent action in one state with potentially violent consequences in others also needs explicit attention. Whatever theories of environmental causes of violence are formulated, they need to incorporate this more complete understanding of cross-boundary linkages to adequately grapple with the complexities of contemporary processes.

The question for this chapter is how to conceptualize these themes in a way that links questions of ecology and sustainability together with some sensitivity to the geography of the processes in motion, a geography that is an important part of international politics but, as the previous two chapters have shown, one that is frequently ignored in the specifications of political danger. Three interconnected conceptualizations grappling with these matters are discussed in this chapter to elucidate both the complexity of ecological linkages over distances and the key theme of interconnectedness in contemporary understandings of environmental change. In doing so, it is suggested that the geopolitical vocabulary of security studies is inadequate for understanding the ecological processes underlying what is supposed-ly an environmental crisis that may lead to political conflict.

ECOLOGICAL SHADOWS

The term ecological shadow first came into prominent usage before the Earth Summit in Rio de Janerio when it appeared in Jim MacNeill, Peter Winsemius, and Taizo Yakushiji's book *Beyond Interdependence*. Concerned with the interconnections across boundaries, the authors suggested that industrialized states

> draw upon the ecological capital of all other nations to provide food for their populations, energy and materials for their economies, and even land, air, and water to assimilate their waste by-products. This ecological capital, which may be found thousands of miles from the regions in which it is used, forms the "shadow ecology" of an economy. . . . In essence, the ecological shadow of a country is the environmental resources it draws from other countries and the global commons.[7]

If the state that draws resources from elsewhere does not in some way ensure the sustainability of the resource base that it draws upon, then its extraction of resources causes a shadow to fall over the ecology of another state. Similar terms can be used to describe the consequences for global commons, be they air, water, or oceanic fish.

The initial formulation did not develop the concept in much detail or use it with analytical precision, although the term itself is very evocative. Peter Dauvergne has elaborated this concept to elucidate the complex social and ecological practices that have caused rapid deforestation in Southeast Asia. He develops the idea, arguing that a shadow is more than merely the aggregate of trees, soil, minerals, and air or an area that is destroyed; it should also include the price paid for this destruction and the related impacts on resource management practices. If a country extracts resources but pays a price that is high enough to ensure sustainable management of the resource, the long-term deleterious effects are less harmful than if the resources are taken without payment for their replenishment.

> the ecological shadow of a national economy is the aggregate environmental impact on resources outside a country's territory of three sets of factors: government policies and practices, especially ODA and loans; corporate conduct, investments, technology transfers, and purchasing

and distribution patterns; and trade, including export and consumer prices, amount and "type" of consumption, and import barriers.[8]

Dauvergne emphasizes that such shadows are a product of many social processes and decisions rather than a deliberate or coordinated plan. Ecological shadows change as consumption practices change and as different commodities are traded. Historical context is important. Shadows are the cumulative result of a global capitalist economy, the attitudes and practices legitimized by such an economy, and the combination of numerous economic and political activities that have transborder effects. States that are highly industrialized but lack a large resource base are likely to have larger ecological shadows due to the necessity to import raw materials, timber, and food. Nonetheless, for analytical purposes it is important to disaggregate shadows into ones caused by particular commodities. Japan's shadow of tropical timber is the primary concern of Dauvergne's analysis.

Japan makes an especially good example of the processes at work in ecological shadows because its industrial economy requires huge amounts of imported energy as well as timber and other resources. Its complex relationships with Southeast Asia are partly about securing supplies of these things from abroad. In the process, Japanese companies have developed complex trading links with local political and economic elites. Many of these have been interested in quick profits, which logging has often provided. What has frequently not been provided in turn is resource management capability and policies to replenish the forests. Logging has facilitated access to forest areas by building roads into previously inaccessible areas. This has allowed farming and accelerated the destruction caused by forest fires, partly fueled by the waste timber and branches left behind by logging.

But the export of ecologically inappropriate Northern technology is also part of these processes. There is a two-way interaction here, but Dauvergne emphasizes that it is an asymmetrical one that effectively acts to transport the costs of Northern consumption to the South. While it is all too easy to move from such an argument to blaming all the ills of ecological destruction on the sometimes appalling environmental record of Northern-based multinational corporations, Dauvergne is careful to point out that while companies have frequently been less than environmentally responsible, they have often had the active cooperation of local political leaders. Indeed, the most obvious strength of his analysis of the detailed pattern of deforestation in Southeast Asia is precisely where he demonstrates the complex patterns of interconnection between Japanese corporations and the patrimonial economic structures of Indonesia, Malaysia, and the Philippines. The intersection of traditional indigenous patron-client relationships with the new opportunities for short-term wealth creation offered by the post–World War II economic boom in Japan shaped the pattern of deforestation.

In emphasizing this connection, Dauvergne's analysis cuts through the state-territorial lenses in which these matters are often discussed. It is because the cross-boundary linkages take particular forms that destruction follows, not because there is cross-boundary trading. Corporations are responsible in many ways because they could operate in ways that encourage ecologically sustainable timber practices, but local elites are often no more interested in these than company directors in distant cities, either at home or abroad. Understood in these terms, simplistic analyses of Northern domination and Southern victimization are rendered more complex, and the simple attribution of guilt is shifted from geopolitically convenient labels to a challenge to social arrangements in both areas. Such thinking is one of the things that the alarmist geopolitical security narratives by authors such as Robert Kaplan miss. As a result, some of the most important policy options don't get considered. Discussing shadows in their complexity contributes to understanding that environmental security is not a simple matter of "their" destruction of the environment somehow threatening "us" in the North.

Dauvergne's analysis points out that neither the model of Northern corporations as the evil investors in the South nor the Malthusian models of Southern irresponsibility threatening Northern populations fit in the case of Southeast Asian forests. This is mainly because of the corporate structure of the sogo shosha who dominate the timber trade in the Pacific. They have not invested in Southeast Asia because they are primarily trading organizations, rather than companies interested in long-term production for which investments in processing plants are required. As a result, temporary contracts, which can be changed or canceled at short notice as market conditions vary, have been preferred. In summary:

Japan's ecological shadow, Southeast Asian policies, and patron-client networks are interlocked. Japan's ecological shadow of tropical timber constricts Southeast Asian decisions and generates short-term financial incentives for destructive and illegal timber operators and their political, bureaucratic and military patrons. Japanese corporate leaders are generally not members of Southeast Asian patron-client timber networks. But Japanese money is often crucial for the survival of these networks.[9]

The resulting deforestation has seriously damaged ecosystems across the region, destroyed the ecological base of survival for some indigenous peoples, and in part led to the environmental disasters of massive fires and air pollution in the late 1990s in Indonesia.[10]

ECOLOGICAL FOOTPRINTS

Where shadows focus on the interconnections across state boundaries, William Rees and colleagues have asked parallel but slightly different questions using the idea of ecological footprints. Concerned with matters of sustainability and the lack of clarity as to what sustainablility means in most contexts, they have developed a simple concept to use as a heuristic device and planning tool of wide applicability. The premise is that humans and their activities are unavoidably a subsystem of natural processes. While this is obvious, indeed precisely because it is so obvious, its importance is often overlooked in scholarly analysis.[11] Rees suggests that this point, reduced to a given premise in much technical analysis of sustainable development, ought to be rethought. Sustainability suggests a stable atmosphere at the largest scale and environmental resources of other kinds that can be produced indefinitely.

To undertake the analysis of the necessary conditions for sustainable human societies, the footprint concept starts by asking how much land is needed to provide the resources for and absorb the waste products of "a defined human population and economy."[12] The initial illustrative example used in their popular introduction to the subject asks the question of how large an area would have to be included under an imaginary dome surrounding a city to ensure that all the resources its population used came from within the enclosed space under the dome. Including consideration of the land necessary to grow food and the land necessary to either grow timber to supply energy or absorb all the emissions of fossil fuels consumed by transportation, industry, and domestic uses suggests that the ecological "footprint" of the city stretches a long way beyond the administrative boundaries of most city limits. "By definition, the total ecosystem area that is essential to the continued existence of the city is its *de facto* Ecological Footprint on the Earth."[13] For real cities, as opposed to the imaginary heuristic one under the dome, this "footprint" includes the land wherever on Earth it exists to supply the necessary ecological services to keep the city going. "Modern cities and whole countries survive on ecological goods and services appropriated from natural flows or acquired through commercial trade from all over the world. The Ecological Footprint therefore also represents the corresponding population's total 'appropriated carrying capacity.'"[14]

Using aggregated national statistics of energy and food consumption, the analysis of this appropriated carrying capacity can be fairly easily done for whole states and hence, when divided by the size of population, for individual human beings. The illustration used by Wackernagel and Rees is the area of land needed to support the average Canadian. Calculating the amount of land to grow food and absorb the carbon fuels used by this individual suggests that each Canadian needs an area of nearly 4.3 hectares or about three city blocks. Calculating on the basis of the land needed to grow crops to provide fuel, rather than simply providing "sinks" for fossil fuel emissions, so that there is no net impact on atmospheric change, makes this individual footprint larger. The statistics for the average American suggest a slightly larger footprint given the higher consumption of energy and other ecological "services." Cross-national comparisons suggest that the average ecological footprint for Americans is 5.1 hectares, for Indians 0.4 hectare, and for the average world citizen 1.8 hectares.[15]

Translated into analyses of the areas needed to provide for the consumption of either a single city or a whole country, the figures are also very suggestive. In the case of the Netherlands, the population needs an area fourteen times the size of the existing state to support its consumption in a sustainable way. Effectively, what large cities or advanced industrial states do is import carrying capacity by using resources from outside their boundaries, both as a direct result of trade that brings resource commodities from afar, and as an indirect result of unmonitored ecological flows. In short:

Northern urbanites, wherever they are, are now dependent on the carbon sink, global heat transfer, and climate stabilization functions of tropical forests. There are many variations on this theme, from drift-net fishing to ozone depletion, each involving open access to, or shared dependency on, some form of valuable natural capital. Pollution externalities and the so-called "common property problem" increasingly erode the integrity of the global commons.[16]

Obviously it is possible to construct statistical measures of numerous footprints and add to the complexity by adding numerous factors and disaggregating statistics according to many scenarios. In making overall evaluations, Rees and Wackernagel take care to be conservative in their assumptions and err on the side of underestimation. Thinking of the planetary ecological system in footprint terms, the calculation of the available productive land that can be appropriated, once 1.5 billion hectares has been set aside for wilderness purposes including the maintenance of biodiversity, suggests that 7.4 billion hectares of ecologically productive land is available. Writing in the mid-1990s, they argued that "Since the beginning of this century the available per capita ecological space on Earth has decreased from between 5 and 6 hectares to only 1.5 hectares."[17] Simple arithmetic then shows that six billion human beings, each with a footprint of 1.8 hectares, cannot live sustainably within the ecological parameters of the planetary ecosystem. The 10.8 billion hectares needed to support this population exceeds the 7.4 billion that are available.

In terms of discussions of global environmental change and matters of security, the alarming conclusion from such an analysis is that the total human footprint is already larger than the available land area to support such activity. This is evident in the growing carbon dioxide levels in the planetary atmosphere that suggest that natural sinks for carbon are not absorbing all the emissions from fossil fuels. But if the rest of the human population aspires to live North American standards of life, with footprints closer to 5 hectares per capita in the twenty-first century, then the planet is simply much too small to support us. In Wackernagel and Rees's choice words, "if everybody on Earth enjoyed the same ecological standards as North Americans, we would require three Earths to satisfy material demand using prevailing technology."[18] Because six billion people each require a per capita footprint of 4 hectares to support Canadian standards, this requires 24 billion hectares where 7.4 billion are available. If ten billion people are on Earth by 2050, each aspiring to use as much energy and other ecological services as Americans did in the 1990s, then 50 billion hectares, or six Earths, will be needed.

This is obviously unsustainable. Such calculations highlight both the gravity of the current situation and the inadequacy of most policy instruments and social science concepts for dealing with current trajectories. They suggest the necessity of rethinking many of the assumptions about modernization and the economic measures of performance used to discuss public policy. They emphasize the ecological significance of the resource flows between states highlighted by the discussion of ecological shadows and the timber trade. Finally, they point to the necessity of looking at practical arrangements that might be taken to reduce both the total footprint of contemporary societies and specifically the most damaging parts of the shadows, which, apart from the destruction of specific ecosystems, when added together, reduce the total available ecologically productive land area of the planet.

But this analysis also points to the interconnections between the different environmental problems in North and South in a way that is often absent in prognostications about global environmental difficulties that suggest a common fate for all of humankind. The rich North has urban problems stemming from high levels of wealth and consumption; those in the South stem from material deprivation and poverty. "However, there is a connection. Much of the industrial countries' wealth came from the exploitation (liquidation) of natural capital, not only within their own territories, but also in their former southern colonies. . . . this appropriation of extra-territorial carrying capacity continues today in the form of commercial trade (as well as natural flows)."[19]

ENVIRONMENTAL SPACE

Recognition of these interconnections and the necessity to reduce the overall impact of human consumption on the natural environment while simultaneously dealing with the interconnected problems of affluence in the North and poverty in the South underlies the efforts of researchers at the Wuppertal Institute in Germany to elaborate policies for "greening the North."[20] As the authors state in their introduction, the links between equity and environment have been noted since the deliberations of the World Commission on Environment and Development, but are frequently forgotten in the discussions of sustainable development and the arcane technical deliberations of international trade negotiations. Specifically, the discussion of sustainability links ecology and equity on three broadly accepted political value judgements—broadly accepted, but not, please note, either completely accepted or interpreted consistently.

The first premise of sustainability is the rights of future generations to live on Earth with access to its resources. Intergenerational equity is key to sustainability; if future generations don't need resources there is no need for the current generation to worry about using them all. The second suggests that natural capital is not substitutable. Promises of an "affluence package" of technological substitutes for natural phenomena are not convincing to many ecological economists by now. The assumption is that natural capital cannot be readily replaced by artificial substitutes. The third is one of intragenerational equity; each human has an equal right to an intact environment. An argument that future generations have a right to environmental resources while the poor currently alive do not is both inconsistent and politically unacceptable.

Using these premises, the Wuppertal team has developed the concept of "environmental space" to refer to "the area that human beings can use in the natural environment without doing lasting harm to essential characteristics. This environmental space is a function of the carrying capacity of eco-systems, the recuperative efficiency of natural resources, and the availability of raw materials. The concept thus expressly recognizes physical 'new limits to growth' . . . resulting from the carrying capacity of eco-systems and the finiteness of natural resources."[21] The emphasis on the novelty of these limits is a reference back to the early 1970s debate about "limits to growth" that suggested that there were physical constraints on naturally available raw materials that would eventually hinder human economic activity and in worst-case scenarios lead to environmental and economic collapse.[22] In the decades since, it has become clear that the limits to the capacity of natural systems to absorb wastes and recuperate from extractions are the most important dimension of planetary limits.

Environmental space understood in these terms, as the ecological use of nature, is then linked to the formulation of principles of sustainability that can be summarized as follows:

1. Utilization of a renewable resource should not be greater than its regeneration rate.
2. Discharge of materials should not be greater than that environment's capacity to absorb them.
3. Utilization of nonrenewable resources should be kept to a minimum. Use should be dependent on creation of a physically and functionally equivalent renewable substitute.
4. The time factor in human intervention should be in balance with that of natural processes: the decomposition of waste or the regeneration rates of renewable raw materials and ecosystems.[23]

Inevitably, the questions of what risks to ecosystems ought to be acceptable are raised. The answer is only low-level risks because of the precautionary principle and the insistence on protecting the weak people in a society. But the inadequacy of comprehensive knowledge of ecosystems is compelling; natural science cannot give economists precise knowledge of the thresholds or limits of complex natural systems. At best, risks can only be minimized.

But with limited environmental space available, the coupling of Northern concerns for global environmental degradation with Southern demands for justice requires that justice be decoupled from conventional understandings of development. The assumption in the Wuppertal study is that long-established patterns of structural power have constructed and perpetuate the injustices of the world, but the solutions are not necessarily to be found in redistribution of wealth. Rather the long-term future of humanity requires an "organization of one's own behavior so that no-one else is systematically deprived of their rightful due. The environmental space available for a society's use is thus calculated on the basis of both the ecological limits and other societies' claims to utilization."[24] The simplest and most compelling guideline for reorganizing societies in line with sustainability and equity remains one of per capita equality. However, recognizing the vastly different ecological conditions in various parts of the world, the study suggests that similar rights should exist for some raw materials whose throughput will have to be varied by region in making calculations of global totals.[25]

The measurement problems of these ideas are daunting. Simple monitoring of particular substances may miss important chemicals or environmental processes and hence needs to be complemented by measures that focus on resource extractions as a measure of overall throughput in ecological systems. Sustainable economies ought to operate with a minimum of nonbiotic extractions of resources so total throughput of material is a key measure. Primary energy consumption is also an essential indicator, as is water use, which is the largest single material throughput for most human systems. Measures of land use that assess its differential impact on natural systems are also necessary but have to encompass an indication of the constraints they put on natural adaptability, too. Coupled to measures of some key emissions, these kinds of indicators offer a baseline for evaluations of environmental space. Some of these calculations and measurements become very complex; the simplicity of the footprint calculations is missing, but as comprehensive management tools the Wuppertal measures do focus on the essential dynamics of human ecological systems.

Discussing these matters in terms of the overall impact of the rich North on the poor South is, as noted earlier in the consideration of shadows, a difficult matter. While toxic waste exports to Africa from Germany may be an ecological crime, the case of Japanese-driven forest destruction in Indonesia is more complex. "Things become more difficult in evaluating imports of raw materials from developing countries. Most of us do not usually know whether mining metals there destroys landscapes or not. With imports of flowers, animal feed, and citrus fruits, are the fertile soils used by crops being exhausted and poisoned by pesticides? Are they being taken away from the poor who must then try to survive in areas of minimal productivity or jungles?"[26] While the answer is usually that details of these processes are hard to come by, nonetheless tracing trading links, as Dauvergne has done in the case of Japanese–Southeast Asia timber-trading patterns, can answer some of the most important questions. The ENCOP case studies, discussed in chapter 3, also offer some answers, as does the growing international network of activists linking matters of environmental change and degradation to questions of human rights and environmental justice.[27]

GLOBAL SPACE: COMMONS AND RUCKSACKS

But when it comes to the big questions of responsibility for global warming and ozone-depleting substances, the appropriation of global environmental space is more obvious even if the solutions to the injustices are especially fraught in the international arena. Here the problem is squarely one of what counts and who does the count-ing.[28] The strength of the Wuppertal analysis is in part its attempt to think past the logjams of mutual recrimination that so often clutter up international negotiations of climate change and related matters. The details of these negotiations are beyond the scope of this book, but the general principles invoked in the negotiation processes be-tween North and South need some attention primarily because they are debates about the appropriate historical and geographical prem-ises for discussion.[29]

The arguments about carbon dioxide emissions are especially pointed when discussing questions of justice and the appropriate counting methods for evaluating responsibility for gas emissions and negotiating agreements to reduce climate change. As noted in the dis-cussion of footprints, the average resident of the North uses much more energy than the average citizen of most Southern states. Looked at in historical perspective, the industrialized states have contributed even more to the total emissions. Their historical use of fossil fuels has been much larger than in former colonies or other states in the South. Some greenhouse gases have considerable residence times in the atmosphere, so such reasoning makes questions of shared respon-sibilities important. Thinking of these themes in terms of strategies of collective prevention of future harm is especially important given the obvious long-term benefits of greenhouse gas reductions, but until some formulas for allocating the responsibilities are agreed upon, progress in international regime construction is especially difficult.[30]

But straight equivalents between emissions for different purposes also occludes the relative importance of different human activities. A decade ago this was pointedly argued by the Centre for Science and Environment in New Delhi, which drew attention to the fact that methane emissions from Southern peasant subsistence plots were being equated with emissions from factories making luxury consumer items for Northern elites in some calculations of the relative respon-sibility for gas emissions.[31] Such arguments continue to be part of the international negotiation processes on protocols for greenhouse gas emissions. In terms of justice, an equation of "survival" emissions with "luxury" emissions is bound to raise the wrath of Southern negotiators.[32]

The additional geographical factor in these concerns is the matter of environmental sinks for pollutants in general and for greenhouse gases in particular. While the oceans are a global commons in this sense, and they do contribute to absorbing carbon dioxide, the question of how to count terrestrial sinks is more complicated. If forests are counted as a global sink and effectively averaged per capita for the whole population, then international attention to the state of forests is inevitable. If, however, individual states claim that forests are national sinks and that these should be counted against total national production in calculation of net greenhouse contributions, then a different set of negotiation positions emerges. Brazil or Russia can claim that their emissions are irrelevant to global calculations because of their large forests. Singapore's politicians will likely object to such arguments. Likewise, it is much cheaper to plant forests in tropical areas, and Northern states are tempted to arrange debt for nature swaps to gain forest environmental space in the South to count against their industrial emissions. The geographical entities used to divide up environmental space are fundamental to how equity is conceptualized and responsibility allocated. Whose security is inevitably a matter of where the consequences fall, but where is a matter of politically constructed geographical categories.

A further geographical consideration in these discussions is the fact that greenhouse gas–induced climate change may impact more severely on Southern states because of the likelihood of increased severity of tropical storms and more extreme droughts and because of the much larger proportion of Southern populations directly dependent on agricultural activities.[33] Disruptions of natural cycles make these people directly vulnerable. In the North, where agriculture is only a small part of national economies and where purchasing power allows for imports to substitute for domestic crop failures, climate has relatively less impact. This connects directly to the points made in chapter 3 about the likelihood of conflict being caused by environmental change. Some of the most vulnerable peoples are precisely those in areas of environmental stress identified by the ENCOP analysis.[34]

In addition to these dimensions of global environmental politics, there is also the matter of environmental despoliation in the South as a result of Northern imports. In the Wuppertal Institute's terms, this is an environmental "rucksack" carried by the consumers in the North. Although the figures are again subject to all the limits of environmental data generalization, the clear suggestion is that, in the case of Germany's food imports from the South, soil erosion there is a hidden environmental cost of German consumption. But worse than this is their estimation that the erosion in the South is on average much worse than that in Europe. Thus imported food has a much bigger overall impact on the planet's ecology than eating food sourced locally. Degradation caused by mining and mineral exports adds to the rucksack carried by Northern consumers. Again, lax or no state regulations in the South may mean that the minerals cause much more damage than if they were mined in Germany. Part of the reason that they are not is simply that easily accessible sources of minerals in Europe have long since been mined. Supplies have to come from further afield as resources close to home are exhausted. In terms of environmental space, all this suggests once again the importance of understanding the cross-boundary effects of environmental activities and the necessity of tracing commodity chains in the international economy to understand the importance of consumption decisions in one place for ecological and social impacts elsewhere.[35]

To these concerns can be added the reduction of biodiversity and the gene pool that is accelerated by green revolution technologies and more recently by genetic engineering. The questions in the international biodiversity convention and World Trade Organization discussions of trade-related intellectual property rights are partly about national resources versus global interests, but they are also about who controls the ecological knowledge and who benefits from this control. Discussing these matters in terms of the biodiversity convention, the Wuppertal Institute authors conclude, "Even if the convention recognizes the traditional biological and breeding skills of indigenous peoples and local communities for the first time, and also assigns them a central role in sustaining ecological productivity, it is apparent that state and/or inter-state organizations will largely be in control, and profits derived from utilization of biological diversity will be divided between national elites and private companies."[36] National elites may be the beneficiaries of international agreements even at the expense of the indigenous peoples on whose behalf these agreements are ostensibly negotiated. Again, the assumptions of geographically uniform national states cannot be sustained when the details of ecological interconnection are investigated.

ENVIRONMENTAL SECURITY, NORTH AND SOUTH

Continuing to think in the Westphalian terms of autonomous states may be aggravating rather than facilitating matters. The histories of spaces and the policies of territorial states understanding their role as the promotion of development may be the result of, rather than the solution to, numerous matters of environmental insecurity. Two further examples of the complex geographical structures connecting security and ecology in practical ways will further elucidate the argument in this chapter on the importance of thinking hard about the specifics of which people and whose environments are being secured and how and where.

Some of the most drastic inequalities and some of the worst environmental degradation coexisted in apartheid South Africa. South Africa's combination of affluence and poverty, extraordinary natural beauty and degraded landscapes, commercial farms and subsistence plots poses an especially interesting case for the analysis of global politics and environmental security. Violence in South Africa escalated as apartheid ended and power struggles in the townships and homelands for access to limited resources became more extreme as the possibilities for change loomed.[37] To understand this more fully requires an analysis of the larger southern African regional context and its historical development as part of larger global patterns. This is especially useful, as noted in chapter 5, insofar as South Africa's apartheid administration in many ways presents a microcosm of the global system.

The regional geography of development in the area is the first consideration.[38] The combination of European imperialism, capitalism, and military innovation that shaped the political landscape was not unique, but in this area "human interventions were particularly upsetting to hitherto sustainable human communities and biotic systems." More specifically, "the most fertile lands were given over to plantation agriculture in parts of Angola, Zambia, Zimbabwe, Mozambique, Botswana, Swaziland, and South Africa; less fertile areas were devoted to pastoralism in all of the above, plus Namibia; in every case, cash crops displaced food crops, and indigenous flora made way for exotics."[39] To support European lifestyles in arid regions and far from the coast, dams and irrigation systems further disrupted the local ecologies.

These arrangements were organized according to the jurisdictional boundaries of the colonies rather than in a way that maintained natural linkages, incorporating the economic systems into the railway lines that ran to the coast to facilitate exports of agricultural produce and minerals from the rich mines that complemented the agricultural plantations. To this day the tailings piles from the mines on the Witwatersrand remind the visitor to Johannesburg of this history. Rivers often became boundaries separating colonies rather than linkages for ecosystems. The state building that followed decolonization and was reinvigorated by neoliberalism and democratization has further extended these patterns of alienation and "development," and in the process provided a powerful series of nationalist arguments and practical measures, justified in the name of sovereignty, for elites to entrench their power at the expense of the population of the region in general and their own states in particular.

Security for these states has been about assuring access to these mines and farms rather than assuring that development served the local inhabitants and maintained their environmental resources. This process of assuring national sovereignty and development as part of the global economy has produced states that are best understood as predatory. Vale and Swatuk suggest that alternatives have to be sought in unconventional considerations.[40] Discussions of peace parks and the possibilities of cooperation across boundaries link up with attempts to allow dispossessed peoples some control over traditional territories.[41] Connections between local populations and ecotourism might also facilitate cooperation across boundaries. Community control over local resources is a key theme in many attempts to revive rural production, although the assumption that local control alone is enough to ensure sustainable development remains problematic.[42] Cooperative ventures around water projects might also facilitate a broader understanding of security that does not get reduced to competitive struggles to assert national sovereignty or maximize conventional measures of national development to pay external debts.

What is clear in the discussion of contemporary changes in southern Africa is that the environmental dimensions of the situation there are important. But an analysis that links them to matters of ethnicity or the dangers of political instability caused by migration, as in a scenario of environmental degradation leading to refugees leading to conflict, misses the most important structural factors in the political ecology of the region. It also misses the powerful historic role of the apartheid state in constructing these identities and attempting to disrupt the anti-apartheid movement by supporting groups organized on ethnic lines while maintaining the huge discrepancies in access to rural resources that consigned many residents of the "homelands" to poverty and environmental degradation.[43]

The Wuppertal Institute understands that the connection between predatory practices by the poorest states, in part to ensure export earnings to pay foreign debts, is the key to policies for greening the North. To reduce the North's claims on the global environmental space requires that at least some of the environmentally damaging exports to the North are stopped. But given that these are often central to state development strategies in the South, elites there frequently clamor for greater access to the tariff-protected economies of the North. Breaking this vicious circle of underdevelopment and ecological rucksacks is essential to any strategy to think seriously about reducing the global footprint of human activity while simultaneously grappling with poverty in the South.

This approach advocates taking less from the South rather than giving more in the form of aid or assistance. By reducing the damage and dislocation in the South, the overall impact of the North on the world's environment is reduced.[44] This is not a call to end all trade, but rather to think hard about the impact of the overall production systems of Northern consumption. It looks to the hidden ecological impacts of cheap raw material imports from the South and the economic restrictions placed on Southern industries, both by tariffs on industrial goods and the export of inappropriate development models and technologies to the South. This is not only a practical political consideration; it also has the potential future advantage of giving those in the North concerned about global environmental problems the bargaining position of having already moved to fix the worst pollution excesses of overconsumption in the North, and hence provides leverage to bring the elites of the industrializing states in the South into line with global conventions.

> The industrial countries must thus take the initiative—for three reasons. First, they have already seriously damaged the environment during two centuries of industrialization . . . and for the moment they are still the worst polluters. Second, the industrial countries have at their disposal considerably more technical and financial means for the necessary changes than most of the South. Third, their life-style has become a model for countries of the South, which will in all likelihood only reconsider what they and we to date term "development" if they see us taking serious steps towards reducing pollution and limiting our own consumption.[45]

To do these things requires rethinking not only development but security and sovereignty, too. Interconnections across frontiers and the recognition of the unintended consequences of consumption in the distant parts of the planet were what first brought delegates to the Stockholm conference on the human environment in 1972. By now practical answers for dealing with these interconnections are forthcoming.

Some examples of innovations from the Wuppertal Institute show that ecological solutions make practical economic sense when the thinking moves from linear reasoning to a recognition of the circular reasoning of ecology.[46] When Munich was faced with rising nitrate levels in the aquifers that supplied its drinking water and the possibility of an expensive treatment facility, the city initiated a program that enabled farmers to convert to organic production in the watershed. Linking a city marketing operation for organic produce with a modest conversion subsidy, the city provided farmers with the necessary framework to make a living while reducing the application of nitrate fertilizers on their fields that would eventually have percolated into the groundwater and required Munich residents to pay for an expensive water treatment plant. While pesticide and fertilizer producers might not be very happy, everyone else benefited as farmers produced local nutritional produce for city residents.[47]

Consuming local organic produce rather than imports flown in from a great distance also reduces the overall impact on global environmental space by reducing the use of fuel for transportation as well as the often ecologically inappropriate use of agricultural land, water, and pesticides in the South.[48] Shifting from monocultures to more diverse agricultural practices and agroforestry in many Southern environments makes ecological sense in many ways if the complex matters of land tenure and resource access are carefully handled.[49] Denser cities that reduce automobile commuting and numerous innovations with combinations of electricity generation and conservation are also reducing total energy consumption in some parts of the North. However, the continued fascination with such things as sport utility vehicles and large suburban dream homes ensures that overall energy consumption is not being reduced.

RETHINKING ECOGEOGRAPHICAL IMAGINARIES

The list of possible Northern innovations is immense; the point here is that these policies are not usually thought of in terms of security. The lack of dependence on imported oil has long been advocated by American environmentalists as an alternative to military preparations and actions in the Persian Gulf. Decentralized grids and multiple suppliers of electricity increase the reliability of power grids in times of war as well as during "natural" disasters. Coupled to taxation policies that tax raw material inputs but not jobs, and energy rather than income, the possibilities for a transition to a more service-intense economy are considerable. But to do all these things requires a consideration of the ecological connections underpinning the global economy that industrial societies have so far failed to take very seriously.

Nonetheless, some security discourses over the last few decades have considered the interconnections between actions in different states. The recognition of the dangers of security dilemmas on a small planet has gradually led to an understanding of the importance of arms control and more broad considerations of common security in many places. The necessity to forego the claims to absolute sovereignty in environmental terms has also slowly fed into the larger patterns of global governance and international agreements and regimes over the last few decades. Extending these understandings into the larger realms of development and rescuing the discourses of progress from a corporate economic specification of its terms is now part of the international agenda of civil society.[50] Key to all this is the recognition of forms of political responsibility that the national state cannot accommodate and the necessity of thinking seriously about the flows of resources and energy that are the basis of ecology as a scientific endeavor.

Despite all this, prominent articulations of environmental security, in the United States in particular, continue to specify things in terms of national security. In the words of one recent academic summary, "environmental security issues virtually by definition exist only in reference to the interests of a particular country."[51] But as this chapter has illustrated, the interconnections across national frontiers are crucial. Simply adding environment into traditional formulations of security fails to go the additional step of asking what ecological thinking might mean for security itself and how it might challenge the identities rendered secure by the conventional practices of state national security policies.

7

Ecological Metaphors of Security

Our attempts today to establish "opulent and completely happy" states still require not only the cultural, and often physical, extinction of many peoples; they continue to require the extinction of thousands upon thousands of non-human life forms as well. The transformations that have been carried out around the globe in the name of happy states have carried with them massive environmental degradation. If we are to confront the crisis of ecology, we must come to understand the relationships between the disciplinary mechanisms that operate within and through our political communities and the potential destruction of our life support system.

THOM KUEHLS, *BEYOND SOVEREIGN TERRITORY*

Surely we have had enough of a politics of little boxes.

R. B. J. WALKER, "INTERNATIONAL RELATIONS
AND THE CONCEPT OF THE POLITICAL"

BEYOND ENVIRONMENTAL SECURITY

Even a brief scan of the security literature suggests that the conventional understandings of security draw very heavily on a mechanical lexicon. Hence "power" is "projected" or "balanced," "force" is multiplied, geopolitical "equilibrium" is asserted as a desideratum. Precisely designated territorial areas are basic to its mode of representation; the "stability" of these containers is understood as the most important task of security agencies.[1] Security is about violence and the control of these spaces. From this it is an easy extension of the argument to suggest that the environment needs to be controlled and that security agencies are the appropriate locus for this political effort.

The more interesting literature in the environmental security debate has suggested that new understandings of security are necessitated by the global environmental crisis. Here physics metaphors of security and quests for control of spaces are understood as counterproductive. Ann Tickner argues that ecofeminist thinking holds some promise for rethinking security along lines consistent with ecological thinking, not least because of the crucial argument that it is necessary to get beyond mechanical, reductionist, and hierarchical formulations in which security is understood as force to reconceptualize security in new and less destructive ways.[2] In the rest of this chapter, these arguments link up with contemporary attempts to challenge traditional ways of thinking in international relations and with the challenges to the spatial ontology of exclusive stable sovereignties and the metaphors of containers.

Most of the critical literature in the field has expanded the remit of security considerations and looked at new lexicons for security or possible new sources of security threats, but it has not thought through in any detail what rethinking security in specifically ecological terms, using the ecologist's conceptual tools of analysis, might suggest about security as a political concept and policy discourse. Taking the environment in environmental security seriously is a fairly obvious line of theoretical inquiry, but it remains, as Eric Laferrière argues for international relations in general, largely undeveloped.[3]

The argument that follows in this chapter proposes that recent research in ecology leads away from reductionist and mechanistic thinking in ways that can link directly to a global conception of ecopolitics. But to link these concepts to security requires both challenging existing geopolitical frameworks and connecting rethought geographies to specifically human ecologies. It adds an additional critical geopolitics dimension to the debates about the much contested political concept of security. In this framework geopolitical specifications of political spaces and the ecopolitical specification of environments and nature are understood as parts of the same process of enclosure, commodification, and territorial control that is so important in the modern colonial imaginary.

ECOLOGICAL METAPHORS

Ecological metaphors for thinking about security are a veritable intellectual minefield given the persistent power of ideological moves of naturalization. If the social can be rendered natural, then it is beyond political control. The realm of necessity is not the realm of freedom and political choice.[4] Scholars need to tread very carefully here. In addition, the changing historical and varied cultural relations between nature and humanity are also important to any assessment of the significance of ecological metaphors. Historians of science might reflect that Malthus influenced Darwin and that the links between politics and biological science are not exactly new. Continuing from the discussion of ecological connections in chapter 6, this chapter suggests that ecology can now be read as a profoundly subversive way of challenging the taken-for-granted categories of international security.

Some of the most powerful metaphors drawn on by strategic studies and international relations in general use terms that relate directly to the natural world. Notable among the natural metaphors in frequent use are references to wild carnivores who threaten unprepared men. Rousseau's stag hunt offers models for cooperation and defection in alliances when hares provide instant gratification.[5] In the geopolitical tradition, the Darwinian metaphors of states as organisms in a struggle for space led, when linked to doctrines of racial purity, ethnonationalism, and romantic essentializations of nature, to Nazi formulations of the need for lebensraum.[6] Perhaps most obvious is the frequent mention of Thomas Hobbes's state of nature where life is supposedly poor, nasty, brutish, and short. This suggests that security is necessary to protect against a prestate existence that renders humans vulnerable to mutual hostility.[7] In light of the discussion in chapter 5, it is worth pointing out that, in part, it seems this view was constructed as a misreading of the situation of the native inhabitants of North America.[8]

The related assumption that nature is a hostile force requiring political organization, and later technical control, to subdue it is an especially powerful trope that should not be forgotten in discussions of the formulation of security. Individual subjectivity is frequently set against a hostile nature in a struggle for dominance. Nature is thus constructed as an external Other to be dominated and controlled.[9] The assumption that humanity is separate from nature or, as matters are later rendered, from the environment is the crucial ontological move, one that influences much of the thinking in international security about the environment. Coupled to positivist epistemologies and contemporary policy discourses based on the technocratic pursuit of knowledge and control, the idea of hostile nature has considerable ideological power. In Kaplan's terms it is rendered as "nature unchecked."

Given the historical understanding of the environment in terms of environs, or something that surrounds, usually a town, this is consistent with the geopolitical practices of security provision premised on the spatial imagination of domestic community within containers threatened by all manner of external dangers.[10] In international relations, environment is often simply used as a residual category for everything of relevance outside a specific demarcation. Harold and Margaret Sprout drew a clear analytical distinction between an environed unit and the environment in which the unit exists.[11] The matter of concern is the relationships across the boundary between that which is inside the unit and the external environment. The international system is then understood to provide a security environment for specific states.[12] Policy is understood as the processes of responding and adjusting to external environmental factors to maximize states' interests in at least potentially hostile situations.

The connection of ecological metaphor themes and the domination of nature to global politics and the discipline of international relations is thus fairly direct.[13] As earlier chapters have illustrated, the domination of nature is the history of European imperialism and the conquest of most of the rest of the earth in search of human and natural resources to fuel the growth of European power and wealth. Plantation agriculture and the use of slave labor to produce commodities in tropical climates were crucial to the rise of European power. The theme of the primitive barbarian, closer to nature because of a supposed lack of civilizational refinement, goes far back in European thinking and has remained a powerful propaganda theme in contemporary times. Extended to the widespread doctrines of *terra nullius,* such geopolitical thinking has had powerful effects on the appropriation of indigenous peoples' lands and resources through the period of colonialism and ever since.[14]

Where Mother Nature is rendered as feminine in gendered tropes that emphasize phallocentric prerogatives to power and control, international relations has perpetuated relations of domination to the exclusion of women and the continued degradation of nature.[15] As noted above, the arguments about gendered constructions of politics and nature go an important step further. The emergence of mechanistic metaphors of the universe and the collapse of the organic assumptions of earlier times in Europe loosely coincide with the construction of social contract theories of individuals understood as atomistic economic actors in a mechanistic chaos that is then referred to as a state of nature. Mechanical representations of nature are not concerned with either the organic interrelationships of earlier understandings or the contemporary ecological theories emphasizing the importance of biological interconnectedness.[16] Nor, as contemporary scholars are increasingly recognizing, are they a very useful epistemological framework for thinking about the complexities of politics.[17]

A further extension of this argument suggests that warfare is not only a long-lasting human invention, but that in some important ways social organization for combat was the first mechanical entity in which predictability and control were centralized and made effective over large distances.[18] One can then continue by arguing that warfare has traditionally been destructive not only of people and their property but of environment itself, both as a deliberate tactic and as a consequence of the destruction and overuse of environmental resources in the processes of war preparations and in supplying armies in combat.[19] Thus arguments for reformulations of the political and dualistic hierarchies of science and politics link up with matters of militarism and the social organization of warfare.[20] Most important for the argument in this chapter, as will be discussed later, the study of ecology and its critiques of resources management has also moved beyond simplistic mechanistic notions of causation to more complex understandings of ecosystems and the biosphere.[21]

The crucial point about invoking ecological metaphors, rather than the biological and organismic ones discussed earlier, is that the object of analysis is not an individual animal or plant but rather a complex of interconnected entities. Extending this shift in focus might be seen as analogous to moving the focus from national to common security. This has already been accomplished in some of the environmental security literature. But the importance of thinking in terms of ecosystems goes further. Ecosystems are understood as matters for the relatively long term and as adaptive systems rather than as a given entity that could easily be conceptualized as in need of securing by practices of violence. While it is easy, although not necessarily helpful, to argue that environmental matters should be a concern understood in terms of security, it is far less clear what security rethought in the light of ecological theory might actually mean.

ECOLOGICAL THEORY 1: ECOSYSTEMS

Twenty-five years ago Barry Commoner encapsulated ecological principles for popular consumption into four basic "laws of ecology."[22] While these are obviously oversimplifications of the complexities of ecology, they are a useful starting point for an investigation into ecological principles. The first law, "everything is connected to everything else," points to the theme of interconnectedness and the impossibility of complete isolation on a small planet; flexible cybernetic systems of feedback function to determine the particular conditions of an ecosystem at any point. The second law, "everything must go somewhere," suggests that there is no such thing as waste; all products of biological processes go somewhere—they don't go away. The third law, "nature knows best," states that human knowledge of ecosystem processes is usually rudimentary and that human attempts to "improve" nature are likely to be detrimental to ecosystems. The final law, "there is no such thing as a free lunch," borrowed from economics, declares that a price is paid for everything and that this is also the case for ecology; energy and materials are used and changed in all ecosystem processes.

Ecology textbooks conventionally discuss ecosystems as the most basic concepts of ecology.[23] The study of ecology focuses on the interactions between biotic communities of organisms (animals, plants, and microorganisms) and the nonliving (abiotic) environment that is in part shaped by the biota. Biotic components include both "primary producers," which use energy sources (mainly solar) to make complex molecules from simpler ones available in the environment, and those that eat primary producers to gain access to their stored energy. A specific ecosystem can, in theory, support a certain-sized community of organisms, its carrying capacity determined by the availability of energy as a result of the primary producers that in turn depend on adequate supplies of nutrients and water. There have been frequent assumptions that ecosystems are basically in a state of equilibrium if undisturbed. Populations of particular species should fluctuate, depending on climate and predator-prey relationships that vary cyclically, within particular ranges up to the carrying capacity in a particular habitat. Good resource management practices, in theory, ought to allow the harvesting of biota with a sustained yield if care is taken to make sure that the carrying capacity is not exceeded.

Human disruptions in the form of burning and clearing for agriculture or other purposes, direct destruction through hunting and fishing, and indirect damage through pollution, as a by-product of many activities, either cause stress on ecosystems or, in many cases, destroy them. Environmental change disrupts assumptions of equilibrium in ecosystems. New habitats are usually colonized by plant species in a predictable sequence called a succession. Subsequent species may grow in size from shrubs to trees and establish a fairly stable "climax" system. As plant species change through the succession sequence, food and shelter opportunities for animals and birds likewise vary so the ecosystem adapts dependent on the relative mixes of different species. Where major disruptions such as fire or clearing happen, similar "secondary successions" often occur without, recent thinking emphasizes, a guarantee that the disrupted ecosystem will return to its initial climax state. Long-term trends also emphasize that stable systems are not as common as formerly thought.

From these formulations it is relatively easy to draw analogies to historical states as entities in flux with changing internal compositions and moments of change, collapse, and regrowth. While ecosystems are not competing directly with one another, nonetheless the analogy holds, up to a point, with threatening phenomena understood as external encroachments. The language is revealing: ecological terms, including "invasion" and "succession," are obviously similar to terminology used for territorial models of states. The boundaries are porous; influences cross these imaginary demarcations. Unlike the fraught biological models of animal territoriality, the metaphors do not include battles over specific areas of land. While ecosystems should not be anthropomorphized, as many animal communities so frequently are, when looked at in isolation their long-term survival could be analogized to states with a reasonable degree of precision precisely because the same basic conceptual assumptions are in play in describing both: semipermanent boundaries and a relatively stable mix of internal components.

But the point about disruptions and degradation is much more important; states, too, are transitory and subject to both internal and external disruptions that change their structure and composition.

ECOLOGICAL THEORY 2: COMPLEXITY, INTEGRITY, AND HEALTH

Where major depletions of minerals or organic nutrients occur or where the soil ecology is eroded or disrupted in other ways, as sometimes happens when forests are cleared, succession may not occur quickly, if at all. In many cases human activities have affected animal species, in particular by disrupting their habitats. Often complex ecosystems, with larger areas and with numerous species or greater "biological diversity," are more flexible in their responses to change than smaller, more limited areas, as they contain greater possibilities for adaptation to the new circumstances. But there are important exceptions to this rule as many tropical rainforests suggest. Because of the complex interdependence of these systems, which are often growing on very degraded soils, when they are cleared they cannot recover.

Recent research in ecology has suggested that at least Commoner's third law, "nature knows best," is flawed in terms of how its role in conservation is commonly understood. Insofar as it suggests that disrupted ecosystems will eventually naturally recover to a situation where a climax state reasserts itself, the logic of this argument is inadequate, both because assumptions of the necessary ubiquity of climax successions are false and because disruptions by human action are so widespread as to often call into question the possibilities of "leaving nature alone" to recover to a natural state.[24] Beyond this Daniel Botkin in particular has argued that ecology needs further rethinking because of its failure to depart from overly mechanistic assumptions, mathematical models of oversimplified ecosystem models, and nonscientific premises of stable and equilibrium natural states as the benchmark for evaluating environmental change and resource management options.

When ecosystems are understood as more variable and less predictable systems that are open rather than closed, and hence vulnerable to interactions on larger scales, it suggests that harmonious assumptions and equilibrium states are not appropriate premises for resources management. The lack of predictability and the multiplicity of operant factors suggest the need for incorporating much complexity in dealing with natural fecundity. Organismic metaphors and discussions on ecosystem health based on equilibrium assumptions are not adequate to consider change and long-term disruptions.[25] Most crucially, Botkin points to the necessity of deciding what kind of nature "we" want, what is to be conserved, and what managed to produce what effect. Clearly we need far more thoughtful, long-term research on ecological systems because the simplistic models of stable ecosystems are not adequate for understanding or useful tools for ecosystem conservation and sustainable management.

The drastic disruptions of ecologies by the European colonial processes of the last millennium also suggest that simply leaving nature alone to recover is impossible given the scale of the disruptions that have already occurred.[26] Further, this point emphasizes that in the case of center-periphery conflicts the assumptions that tribal peoples live in relatively untouched environments that have to be protected from external depredations does not adequately explain the conditions of many peoples struggling over environmental issues. In the case of the resistance in the Narmada valley, Amita Baviskar warns of the dangers of intellectuals romanticizing traditional ways of life in precisely this way, not only because it misrepresents the struggles of local peoples, but because it may also lead to political decisions that are inappropriate for either the peoples or their modes of resistance.[27]

These criticisms of traditional ecological research, especially of many resource and conservation management practices, on the basis of problems within their widely accepted ontological premises, make the simple appropriation of ecological thinking and its applicability to reformulating state-based notions of security a dubious procedure that is less likely to be helpful in rethinking security than it seems at first glance. The analogies, however, suggest some facets of concern to international security, not least long-term changes and unpredictable contingencies. The collapse of states is not unheard of, and the disappearance of the conditions for the reestablishment of state rule is a concern of scholars who have discerned a phenomenon of failed states in the contemporary world. Suggestions that new political arrangements unlike the traditional state may be necessary are also under discussion, but these analogies are stretching the point somewhat.[28]

ECOLOGICAL THEORY 3: THE BIOSPHERE

At the largest scale, the sum of all ecosystems, these matters are often discussed in terms of the biosphere, referring loosely to all life on the planet and the physical components of air, water, and rocky substrate that interact with soil, vegetation, and animal life. Research at this scale encompasses discussions of planetary change on the longest timescale of continental drift, down to more modest matters of ice ages and the long-term cycles of climate change and atmospheric composition, as well as periodic catastrophic events like volcanic eruptions and massive meteor impacts.

The focus on the biosphere has been the theme in discussions of James Lovelock's Gaia hypothesis where life on Earth itself has been understood as a single system.[29] While controversial when interpreted as a scientifically established principle, and it is important to emphasize Lovelock's formulation of the concept as a hypothesis primarily of use as a heuristic device, this framework for thinking emphasizes the interconnection of the planet's parts, including life. The most obvious feature of the hypothesis is that the atmosphere of the planet is in a remarkably unstable state, with reactive gases like oxygen and methane present in ways that suggest a powerful biochemical system maintaining a chemical imbalance that would quickly change if left to its own devices. Lovelock's suggestion is that there are a series of homeostatic regulating mechanisms that are quite sensitive to changes in temperature, ultraviolet radiation, and other important properties for life. This is not a teleological construction, despite its frequent interpretation in these terms.[30]

Crucial for the argument in this chapter is the potential for life to modify the global habitat. Humans seem to be doing it on a large scale and accelerating the process rapidly through the last century. In the biosphere, by far the largest contemporary biological changes are driven by human activity.[31] The dangers of anthropogenic degradation are, at the worst, that our activities will upset the self-regulating mechanisms at a planetary level, leading to some drastic and accelerating feedback loops that will disrupt the global climate system, and possibly the ozone layer, and hence cause ecological collapses that will kill huge numbers of humans prior to some new system being established. Human extinction is an extreme possibility. In ecosystem terms, a situation where one species is appropriating nearly half of the primary production is potentially highly unstable. Of course, the diversity of ways in which this appropriation is occurring suggests that simple systems models will not be a reliable guide to predicting the likely long-term impacts.

The point of this analysis of the biosphere is to shift the focus to the ontological givens of human existence on a finite planet. The Gaia hypothesis forces attention on a simple point that is frequently obscured by ontologies that draw a distinction between human and non-human life: we are not separate from nature, nor is it useful to think of humanity as on planet Earth. We are planet Earth, and as a species we have the ability to change the ecology of the planet. To borrow again from Michael Dillon's epigraph to the introduction of this book, we are a matter of both nature and civilization, and civilization does things to nature in the process of doing things to itself. This line of argument focuses again on the crucial importance of the ontological categories of inside and outside in the formulation of international relations where the discourses of danger so frequently reside.

Perhaps this is most clearly seen in the commonly understood formulations of the relations between the environment and human economic activity. If these are understood as separate boxes with interactions between them, one gets one policy picture. But if, as a focus on the biosphere makes clear, the economy is understood as within the biosphere, rather than as a separate system with some interactions with the environment, and the growth of the economy is absorbing ever more of the biospheric primary production, or natural capital, then its limits and disruptions become much clearer.[32] This brings the argument back to the questions of how one thinks about security, what is secured by this system, and the possibilities for thinking politically about ecology at the biospheric scale.

HUMAN ECOLOGIES

If the analogies of ecosystems with states only partly hold, and the focus on the biosphere forces attention to the limits of international relations concepts of global security, then it follows that states may be a poor framework for thinking about ecology. If it also follows that there is some merit in thinking about security in ecological terms, despite the imprecise fit with states, then it is also appropriate to rethink the political entities that can be considered in terms of security. In particular then, some ecological models of human behavior may be especially instructive, in ways that suggest that the point about ecosystems not being closed systems has pedagogic potential. In terms of the flow of energy through human ecosystems, the appropriation of primary production, and the disposal of wastes, the ecological unit is often the biosphere as a whole. But within the biosphere, as earlier chapters have suggested, there is a complex history of ecogeographical change and appropriation that might be explicated helpfully by thinking through human actions in ecological and geographical terms.

Using Dasmann's distinction between "biosphere" and "ecosystem people," Madhav Gadgil rethinks the causes of environmental degradation and the human role in these processes by differentiating between those who live on the basis of local resources, the ecosystem people, and those whose economic power allows them to draw resources from greater distances, the biosphere people.[33] Ecosystem people are the local cultivators who live a subsistence existence drawing sustenance from local biological resources. Trade is usually local, and droughts or disruptions on a large scale are potentially disastrous. So too, as local populations usually understand well, are failures to conserve the ecosystem to ensure its continued fecundity. Deforestation and the degradation of soils or the collapse of populations of local animals can all imperil the survival of ecosystems. Without romanticizing subsistence patterns of life, as a generalization they are usually relatively sustainable when local resource ownership is clearly related to the maintenance of biodiversity.[34] They are not, of course, permanent, but some of them, especially in Asia, have been relatively stable for millennia.

In contrast, biosphere people are able to draw food and other resources from greater distances. Writing with Ramachandra Guha, Gadgil has suggested that the biosphere people might be termed, following ecological parlance, "omnivores," given their propensity to literally eat everything.[35] This category includes nearly all people in the developed states and the elites of the underdeveloped parts of the world. In contemporary urban societies this is effectively from all over the planet as international trade allows the flow of food and other commodities on a global, or biospheric, scale. Coffee for North Americans and Europeans comes from numerous places in the tropics. Wheat is shipped worldwide; oil, cotton, and so on, likewise. Industrial production increasingly operates on a global scale, too, with the especially pollution-prone parts of the process often placed where environmental regulations will not provide difficulties for corporate management.[36]

This geography has a number of consequences that link the themes of the analysis in chapters 4, 5, and 6 together. Most important for Gadgil and Guha's arguments is the simple fact that the elites, who comprise the most important economic components of the omnivore portion of humanity, are capable of distancing themselves from the deleterious effects of their resource extractions and also from the consequences of the disposal of their wastes. This happens at least in part now at a global scale, despite the attempts to curtail the international trade in hazardous wastes in such arrangements as the Basel Convention. But the metageographies of states as containers of political life cannot adequately capture these processes, which, while they might be partly understood in terms of globalization, have, as the last three chapters have shown, more specific geographies than that general category can elucidate.

This is especially clear when Gadgil and Guha add in the third ecological category of humanity, the category of "ecological refugees." These are ecosystem people displaced from their locales by the commercial processes that invade traditional societies' spaces to feed the biosphere people's economic wants. As commons are enclosed and subsistence plots taken over by commercial farming, the displaced people become refugees and move either to the cities or onto marginal lands.[37] There they destroy more fragile forests and grasslands trying to eke out an existence in areas that are unsuitable to agriculture and for which they have neither the knowledge nor the social and economic incentives to use in a sustainable manner. "The process has its roots in the ever-growing resource demands of the biosphere people, and their willingness to permit resource degradation in tracts outside their domain of concern."[38]

The extraction of resources from ever more remote locations brings with it the further displacement of ecosystem people, many of whom end up in the slums of rapidly growing urban centers where they in turn become impoverished dependents on the biospheric economic system. These are, of course, precisely the people that Gunther Baechler and Thomas Homer-Dixon identify as those most directly at risk of environmental scarcity-induced conflicts. They are also the people analogized with dump bears in the Mi'kmaq activist's formulations of these things in chapter 5.

The North has impacts elsewhere in the world; its shadow ecology, or footprint, stretches to great distances but usually remains a shadow to those who use the resources in the cities because the origin of these resources is not known or considered. In terms of discussions of environmental security and the causes of disruptions and potential dangers to Northern security, the important point from these analyses is to focus on the disruptions caused by the consumption of the omnivores in the global processes as a matter for foreign policy consideration. Gadgil and Guha's formulations and footprint analysis try to explain the consequences of current resource consumption patterns and draw attention to the discrepancies in terms of who gets what where and who pays the price for what kind of consumption.

Neither of them is concerned very much with the category of the state, although, as noted in chapter 6, footprint analyses of particular states are instructive. They both point to the importance of understanding things in biospheric terms, not limiting analyses to individual ecosystems or states but understanding the interconnections between things regardless of geographic convention or physical distance. Indeed, precisely the most important point of such thinking for analyses of security is that phenomena are interconnected in ways that state thinking cannot adequately grasp. The assumptions of communities within states and anarchy beyond is not a very useful starting framework for thinking about anything to do with ecology.

Both these ways of viewing the global polity can be enhanced by focusing on the shift from ecosystem people to the growing number of omnivores and ecological refugees. These are increasingly dependent on the urban economy directly for their livelihood and the provision of their daily needs. While the twentieth century was designated by various terms, such as the nuclear age, it was also the century when humanity effectively became an urban species. While there are numerous measurement difficulties with defining the urban, clearly the world's population is no longer predominantly rural. World cities dominate the patterns of the global economy and cast very long ecological shadows.[39] Many who live outside formal city boundaries still depend on the urban-controlled economy for food, shelter, and livelihood.

THE GLOBAL CITY AND WORLD ENVIRONMENTAL POLITICS

As we become an urban species, it is perhaps appropriate to think about world politics in Warren Magnusson's terms as a problem of urban politics: we are all members of the global city with various municipalities trying to direct its public affairs under the influence of multiple social movements striving to actualize their vision of the good life.[40] The Commission on Global Governance takes this language even further, talking of a global neighborhood.[41] Daniel Deudney has even suggested thinking in terms of a global village and an earth religion to reimagine an intergenerational public and environmental sovereignty, but Magnusson's metaphor is more analytically useful here for extending the question of world politics and specifically how one might now think of ecological security in a global city.[42]

Magnusson's suggestion that thinking about world politics as a matter of urban politics within which social movements clash is a useful alternative framework. It directly challenges the conventional assumption of states as political containers and also emphasizes the importance of understanding politics as dynamic rather than as a series of insides and outsides premised on fixed containers. It suggests that statist political projects and capitalism itself can be better understood as social movements, albeit relatively successful ones. The consequences for thinking about politics in general and security in particular in these terms are profound. Among other things, environmentalism then becomes another series of, albeit highly fractured, movements competing in the political spaces of multiple municipalities and, importantly, across their boundaries.[43]

Understanding world politics as municipal politics with clashing social movements not limited to operating in a single municipality also suggests analogies that might be useful in thinking more intelligently about environmental politics. Taking Gadgil's discussion of the patterns of political power that allow elites to distance themselves from the ecological consequences of their actions, both in terms of their resource exploitation activities and their waste disposal concerns, it is possible to read much of environmental politics as matters of siting decisions and "not in my back yard" (NIMBY).[44]

If we understand this on the global scale, then it is possible to read environmental politics as the siting decisions of global resource flows and industrial production. Not only are the dirtiest of public facilities often sited outside the North, but those areas are also where unsafe pesticides are widely used, unsustainable logging practices encouraged to support export markets, and mining practices used to produce minerals in ways that would be unacceptable in many Northern municipalities. The siting decisions and the fights by local communities to resist waste facilities are now increasingly considered in terms of environmental justice.[45]

Just as the politics of siting is also increasingly understood as a matter of social justice at an urban scale, where the poor and marginalized usually end up with waste dumps and other noxious facilities in their neighborhoods, so too at a larger scale there is, it seems, a similar pattern of siting in the areas of least political resistance.[46] Nuclear weapons testing ends up on the islands of the Pacific, uranium mining on the lands of the indigenous peoples of North America. Nuclear waste repositories are not proposed for affluent urban areas but for remote hinterlands where the relatively politically powerless, who are often the indigenous peoples of the Fourth World, live. Thus the struggles in the Fourth World are overlain with a larger political economy of siting dirty things.[47] Sometimes the dirtiest parts of the North's ecological shadow are most acutely felt at a great distance from where the commodities are consumed. Whether on the small scale of aggregate quarries in Nova Scotia or on the largest scale, as in the tragedy of the Aral Sea where millions of people suffer the ecological consequences of industrial-scale cotton production, or the displacement of forest dwellers in numerous places in the South, these processes are connected with the export of resources to the North.

One advantage of Magnusson's formulation of a single global city is the recognition that there is no place on Earth too distant to be involved in these politics now. All municipalities are involved in the struggle for urban quality of life; whatever the details, in many municipalities forms of environmental resistance are ubiquitous even if not always understood in these terms. Some have powerful influence on the electoral and administrative systems, and active local environmental movements ensure that siting decisions will go in their favor. Impoverished municipalities without substantial tax bases are more likely to get the garbage dumps and polluting factories.[48] Environmental injustice operates at a global scale, and the politics of boycotts and solidarity for activists whose lives are taken as a result of their attempts to maintain the ecological conditions for local survival are an increasingly important part of contemporary political life.

Understanding world politics as urban politics allows a slightly different reading on the environmental security formulations of the South as threatening the North. Perhaps most obvious in Kaplan's stark and evocative prose, the specification of the planet as consisting of tame and wild zones, domesticated zones of order and prosperity in contrast to zones of environmental degradation and population growth–driven chaos, is a powerful trope. There is concern in the environmental security literature that disruptive migrations will imperil the peace and prosperity of the Northern municipalities. As noted in chapter 5, the cover of the *Atlantic Monthly* issue that printed Paul Kennedy's article captured these matters succinctly.[49] Suburban affluence is challenged by the multicultural mob looking over the picket fence. Such is the geopolitical logic of Malthusian fears of migration.

PARKS AND THE COLONIAL IMAGINATION

These themes of urban politics link to Northern visions of parks as crucial to the preservation of nature and to the leisure requirements of city dwellers. If ecotourism is supposedly the solution to some biodiversity problems, then the construction of parks within which the tourist can safely enjoy the natural wonders of diverse ecosystems is too frequently understood to be the appropriate mode of preservation. In the last decades the construction of parks for biodiversity protection has often been undertaken despite a clear understanding that preservation that requires the removal of indigenous inhabitants is unlikely to be a satisfactory method of conservation in the long run.[50]

To construct parks as places that are natural in the sense of not having human inhabitants overlooks the history of colonialism in numerous places and the traditions of constructing game reserves as exclusive domains for hunting by colonial settlers.[51] To make these spaces, the peoples who had gained their livelihoods from these territories were ejected and turned into poachers when they tried to maintain traditional modes of existence. Despite numerous clearly articulated objections to this mode of biodiversity protection, it continues to be promoted, often by international agencies in cooperation with states who are at war with the indigenous peoples on whose lands the parks are created.[52]

Another extension of this argument, in a slightly different vein, is clear in the rapid expansion of golf courses across the Asia-Pacific region in the 1990s. While these are usually understood to be manicured gardens in an urban aesthetic, they have dramatic disruptive effects on the numerous people evicted as well as on the landscape that is physically changed by course construction, the introduction of exotic grass species, and pollution from the widespread use of fertilizers and pesticides.[53] These spaces are then used by elites to further their business and social status enhancement at the cost of the excluded poor in a move that further extends the privatization of working landscapes.

Loosely similar processes can be seen in the expansion of populations into rural areas that depend on urban economies for their livelihood. Telecommuting and the transportation possibilities in courier services allow numerous urban economic activities to occur in remote areas. Coupled to second homes and rural recreational activities such as hiking and skiing, not to mention boating, the rural economies of resource production and farming are often no longer the dominant economic source of income. Conflicts between newcomers and farmers or, as in the case of Kluscap's mountain, between those who value aesthetic considerations and the peace and quiet of rural living and those who value the economic benefits of resource extraction, result in politics that shape the use of land far from the city and accelerate the process of urbanization of the countryside. Ironically, the urban sensibility of environment as recreational space or bucolic bliss competes directly with the traditional rural economy, and in turn sometimes pushes extractive production farther away from urban centers and exacerbates the scale of the urban footprint even as its practitioners try to escape the most direct "negative externalities" of urban existence.[54]

The dangers of continued urban colonial understandings of resources, environment, and nature are directly linked to the construction of the environment as separate from culture, rural as separate from urban, and urban-dwelling, civilized humanity as separate from nature.[55] This line of thinking can be extended by the technological optimists in profoundly worrisome directions where assumptions about technological substitution can be used to argue that environmentalist concerns are irrelevant in the face of technological acumen that can solve any problems.[56] Not that nonrenewable resource supplies may be a problem, but, as Botkin in particular suggests, when extended to the realm of renewable resources these assumptions are highly dangerous because they do not account for the complexity or interconnectedness of natural systems.[57]

Understanding world politics as urban politics emphasizes the importance of thinking through how urban concepts are invoked and used in discussing a separate environment. If the ecologists like Gadgil and Lovelock are correct, there is a basic ontological limitation in this thinking that is related to the conceptualization of politics and science, not only in terms of reductionism and mechanistic thinking that the ecofeminist security line of criticism explores, but in terms of the claim to objectivity being a culturally specific urban knowledge that links to geopolitical specifications of security in terms of the will to power over external environments. Related to the etymology of the term "environment" as that which surrounds, the persistence of a geographical distinction between country (wild zone) and city (tame zone) continues now in terms of threats and new security discourses of danger to the urban residents, threats that ironically sometimes also emanate from mobs in the urban jungle. Inside and outside, danger and safety are multiply coded in this ecogeographical imaginary.

The analogies of these models of biospheric people and world municipal politics could no doubt be further extended, but the points that they suggest are adequate to the limited argument in this chapter: security understood in terms of ecology implies very different understandings once the geopolitical constructions of domestic politics and external threats are confronted by both the geographical dimensions of human ecology and recent conceptual advances in ecological thinking itself. If there are no insides and outsides in a planetary predicament driven by the flows of resources, exploitation of remote ecosystems, and global traffic of wastes, then these arguments further reinforce the basic point that ecological security cannot be understood in the conventional geopolitical parameters of territorial states. But more than that, the whole question of modernity as unsustainable is raised unavoidably by the juxtaposition of security and global environmental change.

Specifically, what this chapter and the two that precede it have called into question is the applicability of traditional formulations of security in discussing the broadened agenda of endangerments that constitutes the post–cold war discussions of global security. More precisely, they have suggested that security is about much more than international rivalries. Indeed, security in ecological terms, as with ecopolitics more generally and a politics that takes the political itself seriously, cannot be understood in the terms of international relations.[58] Disciplinary perspectives beyond political science will, it seems, have to be an integral part of the reformulation of security and part of any project to construct a critical security studies. The possibilities for multiple scholarly perspectives on these issues are the topic of chapter 8, but such a discussion cannot be completely divorced from the necessity to consider once again the entities and identities that are rendered secure. This is the matter for the final chapter.

8

Ecology and Security Studies

We have to refuse again and again, any offers of single vision salvation: these dogmatic commitments would make it much more difficult to accommodate unforeseeable changes, their seeming simplicity would be negated by counterintuitive complications, their rigidity would foreclose precious flexibilities.

VACLAV SMIL, *GLOBAL ECOLOGY*

It is not only the problem of sovereignty that we (another fictional community) need to free ourselves from, but also the problem of political community. In effect, this means finding a way to think about politics in the absence of its defining, constitutive fiction: something far easier to suggest than it will be to effect.

BARRY HINDESS, *DISCOURSES OF POWER*

ECOLOGICAL SECURITY?

Arguably, the most important facet of the recent discussions relating ecology to security is that stability in systems is temporary and that long-term fluctuations are inherent in natural phenomena. The brief discussions of environmental history led to the same conclusion. The dilemma of conservation is that it is premised on preserving that which is changing. Conservation models based on the stability of ecosystems or the possibility of precisely calculating sustainable yields are dubious tools. Given the scale of human activity, an additional consideration is that we have already disrupted and changed ecological systems so drastically that conservation in the traditional wilderness sense is not an option. Relative nondisturbance is necessary in many locations to protect ecosystems and biodiversity. Human impacts on all ecosystems, even the most remote, are now unavoidable if only in the sense that carbon dioxide levels in the atmosphere, ozone depletion, or climate change are affecting plant growth rates and numerous other biophysical processes.

Reflecting on the urban understandings of politics calls for caution in uncritically adopting the conventional goals of political life concerning emancipation understood as urban liberty. If critical security studies is to offer political alternatives to conventional arrangements, then modern urban liberty premised on unsustainable footprints is part of what will have to be rethought. Gadgil and Guha's specification of the human condition in ecogeographical terms challenges the urban assumptions of emancipation and the inevitable benefits of modernization by pointing out that these practices operate to distance the victims of these arrangements. The bourgeois is so termed because they live in towns. Understood as omnivores, their ecogeographical footprints stretch far into the surrounding municipalities. Given the damage done to the biosphere by these modes of production and consumption, cosmopolitan urbanity as the highest human achievement has to be reconsidered in light of the ecological exploitation that is geographically distanced. Critical social theory needs to grapple with these matters in a way that no longer takes for granted the superiority of urban modes of life and the universal claims of the aficionados of globalization of any political hue. Discussions of environmental change cannot ignore questions of social justice, but the ecological critique suggests forcibly that the converse is also the case.[1]

This observation makes the question of what is to be secured especially important. The possibility that the ecological costs of globalizing omnivorous consumption might drastically destabilize the biosphere is the rationale for many invocations to think about environmental security, as well as the related appeals for global environmental management that so worry "global ecology" thinkers like Wolfgang Sachs.[2] While Peter Taylor calls such a program an eco-fascist world order, the World Order Models Project has discussed these matters in terms of eco-imperialism and made the argument that such practices are effectively already in action.[3] Tim Luke's warning that environmentalists often, if sometimes inadvertently, support such projects in their zeal to monitor and encourage managerial responses to political crises extends these observations to once again emphasize the importance of the discursive politics of forms of ecocritique.[4]

From this it is clear that a program of environmental management will have to understand human ecology better than conventional international relations does if world politics in the global city is going to seriously tackle environmental sustainability. Accelerating attempts to manage planet Earth using technocratic, centralized modes of control, whether dressed up in the language of environmental security or not, may simply exacerbate existing trends. The frequent failures of resource management techniques premised on assumptions about stable ecosystems are even more troubling in the case of claims about the necessity of managing the whole planet. Given the inadequacy of many existing techniques, if these practices are to be extended to the scale of the globe, the results are potentially disastrous. In the face of extreme disruption, no comfort can be taken from biospheric thinking or the Gaia hypothesis. As James Lovelock has pointed out, the question for humanity is not just the continued existence of conditions fit for life on the planet. In the face of quite drastic structural change in the biosphere in the past, the climatic conditions have remained within the limits that have assured the overall survival of life—but not necessarily the conditions suitable for contemporary human civilization.

The political dilemma and the irony here is that the political alternative to global managerial efforts, that of political decentralization and local control, which is often posited by green theory, frequently remains in thrall to the same limited political imaginary of the domestic analogy and avoids dealing with the hard questions of coordination by wishing them away in a series of geographical sleights of hand coupled to the rearticulation of the discourses of political idealism.[5] Given that the ecological analyses of biospheric processes and the human ecology discussions of biospheric people suggest both the global scope of processes of disruption and the intrinsic instabilities of ecology, the importance of politics and the inadequacies of international relations to grapple with its complexities is only emphasized in the face of these calls for either global management or radical decentralization.[6]

The widespread failure of the omnivores to acknowledge the consequences of their actions is a crucial part of these concerns, and this responsibility is often obscured by the construction of security in terms of technological and modernist managerial assertions of control within a geopolitical imaginary of states and territorial entities, urbane civilization and primitive wilderness. But as the focus on human ecology demonstrates, nature is not just there anymore; it is also unavoidably here, in part a consequence of human activities, which, although often out of sight to urban residents, cannot remain out of mind in considering matters of world politics and the radical endangerment of human "being" as a result of the practices of securing modern modes of existence.

RETHINKING SECURITY STUDIES

To reiterate the starting point of this book, security does things but is not an entity in its own right. Hence the specifications of danger are crucial to understanding the practices of security; the ability to invoke effective discourses of danger is a political resource that is contained within widely shared geopolitical imaginaries. As the analysis in the latter part of this book has hopefully made clear, once the ontological, and more specifically geographical, presuppositions of security discourse are unraveled from a number of different vantage points, including environmental history, indigenous identity, and ecological theory, the political claims of danger and safety appear in a specifically modern guise. This insight is not itself unique or especially helpful in formulating political responses to the current crisis, but thinking about these categories critically does render what is being taken for granted as that which ought to be secured, problematic.

In a line of argument loosely parallel to the discussions in Rothschild and Williams's work on liberalism and security discussed in chapter 1, Gwyn Prins argues that security studies faces a major rethinking due to the collapse in 1989 of the French Revolution "solution" to political difficulties wherein the state provided some basic security for the population within its bounds in return for the population's incorporation into a geopolitical "hybrid" of the nation-state.[7] The suggestions Prins draws from the events of 1989 are that the narrow focus on military power in realism and the assumption of contract theory have collapsed with the emergence of globalization and the demise of the cold war. While the judgment that 1989 marks such a watershed in terms of political change may be premature, what he implies, but does not develop in detail, is the suggestion that globalization splits the world more irrevocably into haves and have-nots: those connected into the electronic communication circuits and the majority of the global population who have never made a phone call.

Drawing on Bradley Klein's recent analysis of the changing geography of security, we can develop this line of thinking a little further.[8] Klein argues that, at least in North America, security is being privatized in the form of a rapidly growing number of security firms and the increasing phenomenon of gated communities where residents live in a neighborhood of restricted access, complete with guards controlling entrances, checkpoints, and a panoply of security fences and surveillance systems. In the South security is also privatized as guerrilla movements, militias, and mercenary corporations bid on contracts to protect transnational corporations' investments in mines and oil wells. The mobilization of constituencies in ethnic conflict further reduces the claims of states to the monopolization of violence. The theme of the retreat of the state in the provision of security can also be seen in contemporary debates about the increasing role of private corporations in policing and prisons.[9]

While this may not amount to a confirmation of Prins's assertion that 1989 marks the end of the social contract of the nation-state, it does call into question the role of the state in security provision. In particular, Prins's insistence on understanding security in terms of changing social practices and the entities that might be secured as objects for analysis, rather than as taken-for-granted premises for scholarly activity, connects to both the Copenhagen framework and to Keith Krause's summary of the major themes of critical international relations that are relevant to security studies.[10] The affluent enclaves within gated communities in a world of globalized poverty suggest a very different landscape of security than that in neorealist assumptions of states as security actors in the international arena.

Privatized security provision partly undermines the spatial logic of states as the focal point of security discussions from within the state. The geographical assumptions of states as security containers might also be inadequate at the largest geopolitical scale. The crucial question of geographical assumptions underpinning security discourses might then lead to an argument that much of the contemporary confusion about security, whose security, and how to think about these questions can be traced to the early formulation of the problematique of international relations as a field of study in the twentieth century.

Without rehearsing yet again the discussions of international relations in terms of either its "great debates" or "great men," its emergence after the First World War in part as an academic response to the question of why the slaughter occurred is important in understanding what subsequently transpired.[11] The post–Second World War emergence of the discipline as a predominantly American concern is also significant. The American concern with superpower rivalry dominated much of what followed, either directly in terms of the analysis of superpower relationships or indirectly by shaping the logic of inquiry in its terms. Interstate warfare and the need to avoid it between nuclear-armed rivals is obviously an important matter for scholarly inquiry. But in the post–cold war world the assumption that security can be primarily understood in terms of the consequences of interstate rivalry is no longer useful. As research reports from many sources attest, violence and insecurity are widespread contemporary human problems, but they may not be best comprehended within the traditional assumptions of either international relations or strategic studies.

Much of the violence of the last fifty years might be better understood in the geopolitical terms of another function of European, and later American, military power. If the underdeveloped world is looked at more closely, the traditional use of European military forces in the role of imperial administrators and pacifiers may be a more apt geopolitical framework for thinking about contemporary violence. If interventions in such places as Vietnam, the Persian Gulf, Sudan, Afghanistan, Kosovo, and East Timor are understood as imperial ventures, rather than as military actions directly concerned with interstate rivalries, a different picture emerges. There has long been a concern among critics of American policies that the failure to understand indigenous nationalism and anticolonial movements as such, because of the bipolar preoccupations of the cold war geopolitical specifications of the world, caused all sorts of violence that was avoidable. The geopolitical assumptions on which the arguments about security are premised are crucial. Adding the large-scale geographical questions of "security for whom and where?" makes these themes clear.

The historical use of European military force in colonial conquest and pacification cannot easily engage the formulation of international relations understood in the narrow sense of interstate warfare.[12] The assumption of newly independent states as formal and legal equals in the system allows for the consideration of military force only in these terms. But if one understands the use of military force as a hegemonic endeavor or even as an international policing action, then an additional question about international relations as the conceptual infrastructure of liberal hegemony is unavoidable.[13] Insofar as the broadening of security studies is about the new threats of environment, migration, population, drugs, and so on, the violence related to these "dangers" may be better understood in terms of the use of military forces for tasks of occupation and pacification. This requires a different overarching conceptualization of the use of force, as well as a shift from the abstract neorealist assumptions of autonomous power balancing entities as the basic ontological given for security analysis.[14]

Viewed in terms of the history of hegemonic policing actions, the "new" dimensions of security seem much less new. Indeed, the extension of analytical work on security to encompass the broader Copenhagen framework, or even the UNDP "human security" framework, was, in many important respects, anticipated by peace researchers a generation ago. In particular, the work by Johan Galtung three decades ago on imperialism and structural violence suggests that the new agenda of security studies is not all that new.[15] What has changed is the demise of the Soviet Union and the shift of focus after the cold war to other matters. The impossibility of now attributing violence in the South to the machinations of the cold war antagonists demands a different understanding of the world political scene and different understandings of security.

But as this book attests, to any security scholar who has made it this far through its pages, adding geographical thinking to security studies introduces a series of categorizations that are not necessarily congruent with conventional security thinking. If attempts to broaden the ambit of security are already resisted within security studies dominated by one discipline, then the potential for larger academic "turf" fights, should a multiplicity of incongruent disciplinary perspectives be admitted to the discussion, should not be ignored. Securing security is very much a matter of politics within the academy as well as outside.[16] This is not only a matter of accommodating diverse methodological approaches from a variety of disciplines. When contemporary critical thinking and intellectual fields such as anthropology, not to mention forecasts of future directions, are added into the mix, profound questions of who precisely is being secured against what threat by the practices of security studies themselves are inevitable. Security is, as Huysmans, Dillon, Walker, Williams, and others increasingly make clear in a variety of theoretical genres, unavoidably a matter of culture and politics in a very unevenly globalizing world.

GLOBALIZATION AND SECURITY

Understanding the geopolitical specifications that underlie particular insecurities is important. Political science, in its comparative guise as well as in international relations, is often preoccupied by the category of the state and fails to look more closely at which state and where.[17] It also frequently fails to ask questions about nonstate communities and the need to think about institutions appropriate to the task of global security that do not take states for granted as security providers. The importance of demolishing the category of the state, in the face of the inevitable insistence on the question of which state, is worth noting, as is the related point about the inadequacy of many states to supply security in terms of the territorial autonomy that neo-realism has always assumed.[18] This, too, links to the assumption that differentiation within a system of connections is a much better starting place for understanding contemporary international relations than the interaction of autonomous spaces.[19]

This argument is important if globalization is to be taken seriously because it changes the focus of analysis from local autonomy to an understanding of global processes occurring in specific places. Much of the focus on the environment as a source of security threats has been on growing populations that supposedly cause both environmental degradation and migration as peasants move when they can no longer eke out a living on degraded landscapes. The neo-Malthusian fears of population- and migration-induced disruptions voiced by environmental security advocates can easily be read as a rearticulation of the geopolitical dimensions of the limits to growth debate from the early 1970s.[20] Social conflict over scarce resources and identity conflicts may indeed result from these processes as analyses by Thomas Homer-Dixon and others have suggested. The literature discussed in chapters 3 through 6 shows that the most important processes are linked to global patterns of political economy and the spread of commercial agriculture and other forms of development, such as large dams, into landscapes marked by very unequal access to land and water.[21]

As chapters 5 and 6 specifically suggest, among the most important of the norms and practices that are being globalized are the European colonial assumptions of land as *terra nullius* unless it is formally surveyed and "title" granted to settlers. The related cultural assumptions of nature as a resource and land as waste unless cultivated by agriculture underlie these appropriations.[22] Thus, aboriginal ecologies and the subsistence economies that they support are disrupted and often destroyed. Bringing anthropologists into the discussion raises the question of the fate of indigenous populations. Critical anthropology suggests that modernity is the gravest threat to all forms of "security" that numerous indigenous peoples face. A parallel process often occurs with the displacement of peasant farmers and the enclosures of "commons" in the processes of expanding commercial agriculture, the only type of agriculture taken seriously in many technologically defined formulations of food security. The assumption that modernization of agriculture is a process that improves security is contested by numerous campaigns on the part of dispossessed peasants.[23] Ecosystem people only sometimes wish to become environmental refugees.

This line of argument parallels others in the critique of development in the last few decades where the assumption that universal modernity is desirable is challenged and the role of marginal locations is reinterpreted as part of the global processes of commodification and expropriation. In economic terms the debate about the global environment, atmospheric change, and sinks to absorb carbon emissions leads logically to the thinking that emission permits are tradable because everything is ultimately reduced to exchangeable commodities.[24] But in light of postdevelopment critiques, one has to ask whether the South simply becomes a tree-growing zone for large Northern states and corporations trying to find a way to sink their emissions. It is not difficult to construct an eco-imperialism argument suggesting that this is simply more Southern subsidies to Northern consumption.[25] Globalization challenges the idea that states are the only obvious referent object of security, although the theme will be especially tenacious where states are seen as the administrative mechanism that ensures (secures) the financial investments in biodiversity and (violently?) administers the North's carbon sinks in the South.

This question of how one understands responsibility, environments, and the matter of sinks reintroduces Prins's question of whether security studies is understood in terms of a science or a humanity. Science suggests security as a matter of technique and technology, a problem to be managed by the appropriate state mechanisms and researched by global corporations. The Archer Daniel Midlands corporation television advertisements in the late 1990s, narrated by David Brinkley, focus on hunger and want as a security problem in need of corporate technical fixes. The invocation of these themes in terms of food security during the broadcast of the CNN television series on the cold war in the last few months of 1998 dramatically raises issues of definition and response. In the corporate discourse hunger is a technical problem above and beyond politics. But most social science literature has long since shown that famines and related social disruptions are rarely a matter of food shortages overall but of local maldistribution, political power struggles and warfare, market manipulations, and the lack of cash among the poor to purchase food.[26] Is this then a technical problem for corporations or a political issue of social organization?

This raises the larger, general question of technical practices and the potential for global managerial efforts. Who should define the problem and design the practices? In light of the numerous failed development projects and related attempts at social engineering of the last half century, attempts to manage social problems at the global scale using similar modes of control raise great skepticism among those familiar with the literature on development.[27] Some, although not all, of the causes of insecurity and violence are a result of the disruptions caused by the processes of globalization. Put in the terms of the UNDP human security agenda, the point is that attempts to enhance economic security through modernization often undermine food, environmental, and community security among the poor and marginalized. Security across these themes is not necessarily additive. Dilemmas abound and imply that politics will not easily be avoided either in practice or in academic thinking about security.

This is a contrary argument to many of the assumptions in the "enlargement geopolitics" mode of reasoning that suggest that modernization and markets are the answer to the conflicts understood as endemic in areas not yet blessed by the presence of modern liberal states and market economies. The idea that modernization is the answer overlooks the violence inherent in the process, often by ignoring the distant consequences of specific changes. The United Nations concept of human security has the considerable analytical advantage of drawing attention to many of these contradictions.

BEYOND INTERNATIONAL RELATIONS?

The emergence of critical perspectives and the addition of methodological concerns from contemporary social theory challenge the cold war definitions of security in a manner that simultaneously broadens the consideration of security and suggests that there is more to the discussion than considerations of academic turf alone. These critical approaches and the addition of wider concerns from other fields call into question the underlying assumptions of security by asking questions that shift the focus of concern away from states as the only, or most important, actors in national security. More explicitly, as Gwyn Prins notes, whatever the aspirations of its practitioners, security studies remains largely a part of the humanities rather than a science. As such it has to be concerned with history and interpretation, matters that concern security scholars despite their often self-imposed nomenclature in North America as social scientists.

All this suggests that security studies has much to offer in terms of understanding the contemporary world scene, but also that it may be long overdue in cutting itself loose from the strictures of twentieth-century international relations thinking and its state-centric premises. If the assumptions of European interstate war, and subsequently the concerns of bloc rivalries, are understood as historical episodes for which international relations in part acted as a legitimization practice for American, and subsequently Western, policies, then the basic parameters for thinking about security might be much better understood by distancing thought from the categories that we have inherited from the past. But as R. B. J. Walker has repeatedly argued, such distancing will not be easy given the limited categories of thought available.[28]

Security studies might draw intellectual support from cultural studies and postcolonial critiques. It might, as Prins argues, recognize itself in terms of a branch of interpretive humanities. In doing so the task of academic practitioners, or insecurity specialists, to adopt Davis Bobrow's term but not his conclusion,[29] becomes one of engagement in a conversation about the fate of humanity on an endangered planet rather than a task of scientists to "objectively" examine a given series of structural relationships. As such, if it also takes the geographical arguments into account, security studies might understand itself as many geographical selves, as security studies plural, with Asian, European, and American variants in addition to a variety of postcolonial and, crucially perhaps, postdevelopment understandings.[30] Security is not then read as a universal condition, as the American behavioral scientists might like us to do, but understood as a highly contested signifier that invokes numerous specifications of danger and legitimates practices of violence while simultaneously frequently promoting a modern liberal subjectivity on the bases of this violence.

It also invokes the modern political impulse to control and the medical tropes of endangerment and disease, abnormality, and threat that are, as Bobrow unreflectively notes, pervasive in contemporary security culture. Foucault's discussions of security, power, policing, and government have suggested that the political impulse to secure the lives of populations within states is linked to the formulations of all sorts of dangers as well as the questions of social welfare. Thus the metaphors of various wars against everything from polio to drugs are part of the larger popular geopolitical repertoires for engaging state agencies. "'[T]he end of the Cold War' represents not a decisive break with the logic of security, but a realignment that throws into relief problems and sites of contestation that earlier had been less accessible, that is, intelligible."[31]

What threatens is a complex matter of cultural politics. Scripts of nationhood are frequently woven around heroic deeds in wartime. Battles and wars of independence are crucial military events when rendered so by the narratives of nation; founding myths produce shared identities. As noted in previous chapters, similar military metaphors saturate many social phenomena concerned with protection and individual subjects. Medical language is replete with bacterial invasions, therapeutic interventions, and the battle with disease. Immune systems are the patient's first line of defense prior to the use of pharmaceutical weapons. Victims of disease struggle valiantly. The battleground is the body analogized to the body politic of the state. Heroic struggles inscribe both national histories and discussions in hospital wards.[32]

This medical intertext also haunts Davis Bobrow's formulation of the tasks of security practitioners in terms of medical metaphors. And yet the formulations of these threats and the invocations of such themes as the "New World Order" once again work within a spatial imaginary where the knowledgeable clinician both specifies the condition of the patient and acts within a disciplinary spatial apparatus that designates the spaces for treatment, the assignment to the asylum, or the terms of quarantine. As François Debrix argues, borrowing from Foucault's analysis in The Birth of the Clinic, "peacekeeping" in the medical language of such organizations as Médecins sans Frontières is also caught in the normalizing practices of the clinical gaze where the failed state, the site of genocide, or a humanitarian emergency is specified in contrast to the abstraction of the normal state.[33]

Such tropes require the intervention of the medical specialist to diagnose, conduct surgery, organize a quarantine, or specify safe havens. What preventive care or a holistic medical metaphor might convey as an alternative mode of knowing the South remains a useful counterargument to the assumptions of clinical medicine even if the answers are far from easy. If at least some of the "new" threats to the geopolitical order of modernity are fairly directly related to the expansion of the impacts of modernity, and Laurie Garrett's extensive compilation of the circumstances of the emergence of new species-hopping diseases due to the disruptions of tropical ecosystems extends the medical language appropriately here, then the matter of causes rather than cures is especially pressing.[34] This is particularly true given the ecological shadows of "normal" states, where the assumptions of indigenous causes requiring intervention to provide a cure fail to incorporate the cross-boundary responsibilities implicit in the causes of the disruptions in the first place.

None of this is intended to disparage the work of such organizations as Médecins sans Frontières in contemporary crises; but thinking politics without presupposing narratives of normalization, spatial security, intervention, and imperial control is one important task for critical thinking "after the cold war." It is if we are to think in terms that do not reproduce the identities, practices, and politics that cold war security discourses secured. Neither can one innocently argue that such organizations, the mendicant orders of the twentieth century, to borrow Hardt and Negri's formulation, are outside the contemporary practices of power however much they may advertise their practices in antigeopolitical tropes.[35] Detaching these things from the traditional ethnocentrism of international relations is a necessary part of the construction of an international and interdisciplinary security studies; it would certainly seem to be necessary if a cosmopolitan understanding of the dynamics of conflict and the possibilities for thwarting identity politics–based warfare is to be the basis of a new political order.[36]

If this broader critical agenda is accepted, can it be used as a way of rescuing politics and security studies from the geopolitical stipulations of communities living in boxes with violence ultimately arbitrating disagreements? Hopefully it can. Can such thinking also shift the referent objects of security from the state to the human individual, as both the human security agenda and the critical security literature advocate, without presupposing the identity of these individuals as neoliberal citizen consumers embedded in a mesh of claims to universal rights? Can all this be accomplished in recognizing that the social processes currently in motion are interconnected, and in that sense global, but not in a sense that requires management of the poor by the rich and powerful? To do so, such studies will need a much more complicated, geographically diverse world of interconnections as the ontological premise for contemplation and analysis. Violence and insecurity then can be addressed more directly without the intellectual detours through assumptions of interstate violence, territorial states, national security, or even "civilizations" obscuring the social processes in question. International relations and strategic studies it is probably not, but security studies it might still be, albeit in a rather different guise from that bequeathed from the hegemonic practices of the past.

RISK SOCIETY AND GLOBAL THREATS

One line of argument that is especially germane in attempting to link such diverse considerations together is Ulrich Beck's sociological discussion of "risk society."[37] Beck doesn't deal with security in traditional terms at all but in a synthetic approach that suggests numerous facets of security in innovative ways. His contribution suggests some possible avenues for critical security studies at the largest scale.[38] Beck's discussion begins with the argument that conventional dualisms between nature and society are not useful starting points for sociological analyses in the period after industrial modernity. Instead, threats and fears are defined in terms of the "fabricated uncertainties" that civilization itself produces: "risk, danger, side-effects, in-surability, individualization, . . . globalization." Technical matters are now politicized in the context of a "world risk society." This typology of global threats distinguishes between wealth-driven ecological destruction, relating to industrial wastes and such phenomena as "holes" in the ozone layer, and poverty-driven destruction—when peasants annihilate their environment in their fight to escape economic marginalization. A third global danger, the potential of nuclear, biological, and chemical weapons, can be added to those of ecological destruction. In combination, these risks are eroding the logic of conventional technological risk, if only for producing situations in which "hard to manage dangers prevail instead of quantifiable risks."[39]

Understanding the question of threats and insecurities in this framework complicates even more the conventional assumptions that environmental degradation presents a threat to the security of Western states. First of all, risks and threats are not purely objective phenomena, but socially mediated political constructs. This is where identity is connected to security and endangerment; risks relate to social institutions, albeit now at the largest of geopolitical scales.[40] That which can be credibly articulated as being threatened is an identity in need of being secured.[41] But the identity of modernity has, until recently, simply been taken for granted in nearly all of the literature on security. If security is not sustainable without ever-growing efforts to expand state control and ever-increasing abuses of local resources toward supplying an ever-expanding production system, then the resulting disruptions and displacements render numerous constituencies insecure. This is especially the case where military establishments take part in resource extraction processes and hence escalate the potential for direct violence against opposition to displacement.[42]

Second, such disruption affects transstate politics by way of an emergent sphere of political activity linking human rights, environment, gender, and development issues. Discussions of global security are premised on the modern assumptions that the state is the provider of security, that legal systems uphold individual human rights, that the latter have been universalized to provide a benchmark for political conduct globally, and that—implicit to much of the conventional security discourse—modernity has to be extended to the poor and backward parts of the world for the greater benefit of all. There remain grounds for dismissing such formulations as ethnocentric despite the semblance of inclusivity conveyed by numerous appeals to global governance.[43] But if impoverishment and disruption is the unintended consequence for certain localities in the global economic system that is necessary for worldwide modernity, the contradictions become palpably painful.

Third, in a domain of global concern, the legitimacy of technical expertise is of crucial importance if trust is to prevail. If environmental risks are perceived as contingent on larger political processes and beyond technical preoccupations with localized impacts, technical expertise itself may lose its claim to legitimacy in irreversible ways. The theory of "world risk society" investigates the emergence of discourse communities capable of arguing that the long neglected side-effects of industrial production must be henceforth understood to entail risks that can deprive the system of its legitimacy and of its "rational" technocratic controls. In many places, environmental degradation goes hand in hand with political criticisms of imperialistic politics and even with nationalist identity politics bent on protecting the homeland—an important but often overlooked process that provided a political factor in the demise of the Soviet Union and of the Warsaw Treaty bloc in the late 1980s.

But also relevant is the fact that "world risk society" pushes politics beyond the conventional parameters, drawing political constituencies together for boycotts and protests around telegenically mediated instances of corporate perfidy and ecological destruction.[44] These emergent discursive political communities are questioning the technical procedures of expert regulatory agencies and corporations on grounds that the ultimate effects of their habitual practices may lead to unforeseen ecological consequences. In the process, the legitimacy of state environmental experts is becoming highly politicized, and the administration of ecological regulations is being politically contested.

This relates to the politics of transstate movements and to global regimes where interstate treaties may provide at least some loose framework apt to constrain state activities.[45] Claims to expertise in environmental disputes are mobilized by both environmentalists and policy makers during the political bargaining processes of international regime formation.[46] State development experts, pollution experts, medical science, and planning procedures are now all in doubt; the politics of technical expertise can no longer be obfuscated under an unquestioned acceptance of the writ of science. "Security experts" are not immune to these developments. The presentation of environment as a threat is a complex political process, not simply an issue "security experts" can paraphrase to elicit a conveniently adequate policy response. But it is a political process that in at least some ways gets well beyond the politics of sovereign states and the conventional assumptions of territorial identities.

POSTSOVEREIGN WORLD POLITICS

The theme of politics as municipal introduced in chapter 7 also suggests that politics is about discussing matters in ways that deal with practical consequences of local actions and that engage in debates about political life that grapple with the questions of how one ought to live in a world where conventional national and state answers are no longer convincing. In the light of ecological understandings of interconnections, adding the conceptualization of omnivores to the identities in circulation in contemporary urban politics might be a very useful supplement to contemporary political discussions, not least because it shifts the emphasis from cosmopolitan formulations that imply universal displaced populations to the recognition of the importance of understanding that the risk society is a participant in ecogeographical processes with multiple distant consequences.

The other point in Magnusson's argument discussed in chapter 7 that is directly relevant here is that politics in these terms partly avoids the question of sovereignty. He tries to open up new ways of thinking and acting politically that decenter the state and reimagine communities and ethical actions. But if the metaphor of the municipality as a particular forum where clashing social movements struggle to actualize their vision of the good life is extended to the globe, it also evades the collapse of political considerations into what states do or do not do and world politics solely into the questions of relationships between state elites. It does not, however, easily escape the assumptions of administrative units as the site of politics. While this line of argument cannot solve many questions of political theory, it does emphasize the importance of conceptualizing nonstate actions as explicitly political and raises the questions of reimagining what it is to act politically once the geopolitical codes of states are relaxed. In Magnusson's formulation, social movements are understood as concerned with influencing outcomes in particular places without necessarily being interested in directly controlling city halls or state administrations.

Models of ecology and resource flows also suggest an extension of some of the discussions of the ethics of responsibility and possibilities of rethinking geopolitical encounters in contemporary international relations thinking.[47] The responsibility to distant strangers becomes very clear when the ecological language of interconnection and cycling is taken seriously, and likewise when footprints, shadows, or considerations of environmental space are invoked. The question of the Other is also the question of the unseen distant other in the rural hinterlands that supply the commodities that the city-dwelling omnivore demands. The resources must continue flowing to keep the city functioning. The hinterland has to be pacified to ensure the flows continue. But the flows disrupt and destroy indigenous culture whose existence is tied to the functional continuation of the ecosystems within which the resources lie. The threats of mass suicide by the U'wa people in the face of the Shell corporation's development of oil fields in Colombia is just a high-profile demonstration of the current politics of enclosure and displacement that is the consequence of the growing automobile-driven urban footprint.[48] Environmental insecurity does indeed have a very specific geography, but a much more complex one than can be understood by neo-Malthusian assumptions grafted onto state territorial reasoning.

The ecological perspective, especially the understanding of humanity as of the earth rather than on the earth, challenges the assumptions of sovereign knowing premised on human separation and domination of an earth distanced by the violent practices of detached and hence supposedly universal knowledge. Appeals to specific situated knowledge are not the simple answer to these dilemmas. Nor is this an argument for making nature sovereign as Patricia Mische implied in the early debates about these matters; her line of argument evades rather than answers the questions about sovereignty and who decides politically by, as she implies, simply inverting the dichotomy.[49] Focusing on the relationships of power/knowledge in the formulation of the planet as an external object to be manipulated, controlled, and managed once again shows the parallels between others and nature implicated in claims to sovereignty and the prerogatives of national security. As the last few chapters have shown, the domination of nature that so concerns environmentalists is precisely what the geography of the colonial project of modernity constitutes.

In contrast is the obvious importance of understanding postsovereign subjectivities in terms of ecological situatedness, connectedness, and change, rather than as permanent, separate, and external. It is therefore also useful to conceptualize the possibilities for political change by understanding ourselves as in motion, rather than in terms of participating in stable, unchanging geopolitical entities in which, and over which, a mode of sovereign power knowledge rules as the highest human accomplishment.[50] The images of these rethinkings are partly caught in Deleuze and Guattari's discussion of rhizomes and complex webs of horizontal grass rather than in discussing states in terms of metaphors of trees with a single focal point in terms of the trunk.[51] But the dynamic dimension needs an emphasis that the metaphor doesn't necessarily convey.

This decentering of sovereign modes of knowledge is in part necessary because the metaphors of environment that have been used in the past are politically inadequate for the new and more complex understandings of natural systems and of ecological impacts in terms of shadows, footprints, and environmental space, which undermine conventional structuralist-based thinking on international relations and the environment. The importance of interconnections and cycling is obvious to any analysis of ecology. So too are ecological notions of change and adaptation in the face of the vagaries of chance events, a point that classical realists may recognize, even if it appears threatening to the neorealists.

Ecological thinking reinforces the critiques of neorealism from international political economy in that it once again suggests the importance of the flows and transgressions of the boundaries taken for granted in international relations thinking. This is coupled to an insistence on the importance of understanding the entities whose borders are transgressed as temporary. In doing this, the basic assumptions of conventional neoliberal institutional approaches to international relations are directly challenged. So too are their geopolitical specifications of fixed territorial boundaries as the long-term premises for political action. As such, ecological thinking suggests additional important arguments for considering seriously what postsovereign politics is and might become.[52] The persistent political question of who we are is especially forcefully put in these discourses. But it also raises unavoidable questions about the identities that contemporary political arrangements are securing and about what is so important that it becomes necessary to invoke security discourses to protect and so to ensure into the future.

9

Securing What Future?

If the subject of security is the subject of security, it is necessary to ask, first and foremost, how the modern subject is being reconstituted and then to ask what security could possibly mean in relation to it. It is in this context that it is possible to envisage a critical discourse about security, a discourse that engages with contemporary transformations of political life, with emerging accounts of who we might become, and the conditions under which we might become other than we are now without destroying others, ourselves and the planet on which we all live. Where so much recent debate about security has been predicated on the impossible dream of absolute invulnerability (the counterpart of the impossible dream of absolute freedom), a critical engagement with security would envisage it precisely as a condition of being vulnerable to the possibility of being otherwise than one has already become.
R. B. J. WALKER, "THE SUBJECT OF SECURITY"

SECURITY AND IDENTITY

Security is about the future or fears about the future. It is about contemporary dangers but also thwarting potential future dangers. It is about control, certainty, and predictability in an uncertain world, and, in attempting to forestall chance and change, it is frequently a violent practice. It is about maintaining certain collective identities, certain senses of who we are, of who we intend to remain, more than who we intend to become. Security provides narratives of danger as the stimulus to collective action but is much less useful in proposing desirable futures. This has long been the function of other narratives, of nation building, progress, the promise of development, the practice of modernization, and the collective aspirations to modernity. But the problem now is that the aspirations of many to be modern, in its contemporary carboniferous form, are endangering the possibilities for a desirable long-term future for nearly everyone. The modern liberal subject is the autonomous, threatened, knowing, heroic identity, not the postmodern temporarily particular corporeality of multiple, contingent, ecological interconnections.

Security discourses specify the endangered identity. But these are multiply coded and embedded in cultural contexts. To be effective they need to interpellate existing social identities and articulate them to other discourses in circulation and to commonsense geopolitical reasoning. As the Copenhagen authors make clear, an important part of the whole question of security is this ability to specify matters of danger, invoke emergency measures, and gain at least some political adherence to the crisis script. As such, security is a profoundly political process, most obviously so when it works to constrain politics. Dissidents can be silenced most effectively when they are portrayed as aiding the enemy. Resources can be mobilized and solidarities cemented in the face of a common threat. Security in these contexts is not only a matter of state action and formal politics but also a matter of civil society where policies and identities are argued, affirmed, and articulated in the routine quotidian practices of culture.[1]

To be effective in stimulating new thinking, concepts like the United Nations formulation of human security face a series of problems dealing with the legacy of cold war thinking and with contemporary geopolitical understandings of security, which despite the broadening of the security agenda still frequently operate on assumptions of internal secure spaces kept separate from external threats by surveillance and technological acumen. That this assumption should be the starting point for security studies is not surprising. The geopolitical themes of protected inside and threatened outside is taken to its obvious conclusion in recent Hollywood movies (among others, *Deep Impact, Independence Day,* and *Armageddon*) where the earth is saved from external destruction by American ingenuity and, of course, nuclear weapons. This point is not a flippant reference to the distractions of popular entertainment, but rather a serious argument that security is part of the hegemonic understanding of who "we" who are insecure are. And who we are, and what metaphors our political leaders can use to invoke discourses of danger, are unavoidably matters of popular geopolitics and practical geopolitical reasoning.

In discussing the politics of security discourse and its links to population concerns and neo-Malthusianism, Betsy Hartmann quotes Andrew Ross to make the point that in the 1990s, despite a booming economy in the United States, fears of scarcity were a powerful part of the dominant neoliberal discourses of economics.[2] Social scarcity is a powerful assumption in this context, and individual economic insecurities in a fast-changing economic system map onto neoliberal assumptions neatly. Ross makes the case that the coincidence of this neoliberal discourse and the assumptions of environmental scarcity in neo-Malthusianism fit together to support a security discourse that represents the impoverished South as both a natural phenomenon and a threat to American affluence.

Whether as consumer or citizen, and the two identities are especially closely related in the contemporary discourses of neoliberalism as well as in many discourses of human rights, this autonomous knowledgeable being is both the ontological given and the political and economic desideratum in contemporary discussions of security and economy. Social and natural scarcity support each other in these political discourses of danger, nowhere more so than in the stark portrayal of environmentally induced dangers to American bucolic bliss encapsulated in the automobile advertisements that framed Robert Kaplan's prose in the *Atlantic Monthly*. The processes of commercialization have long had within their expanding economic reach a powerful theme of identity production and status aspiration. As Nicholas Xenos makes clear, the rising power of the merchant class in imperial Britain and their diffusion of London fashion into the countryside set in motion a system of status related to the possession of the latest styles and accoutrements of modernity.[3] The newest, the fastest, the most exotic, the latest innovation is a mark of status and affluence, and of cultural accomplishment and aspiration as well, in the lives of those who consume.[4]

But once an item becomes widely dispersed in a population, it no longer has this cachet. Scarcity is key to this and a persistent, powerful assumption that drives desire and consequently consumption. Applied to those who are patently not like us, in terms of their possessing the cultural attributes of affluence or modernity, this assumption leads to the general specification of scarcity as the lot of all people, and hence to the Malthusian concept wherein other cultures are defined by their lack of modernity. Within the affluent "happy states," anxieties about appearances and keeping up with, or preferably getting ahead of, the Joneses play out in ambiguities about status, perpetual economic insecurity, and complex political debates about the role of the state in redistribution and the claims to numerous entitlements and rights.[5]

These anxieties are not removed from the larger question of moral order and the specifications of dangers to the polity. "So state policies that appear at one moment as legitimate and fair because they are universal and thus benefit oneself seem the next moment to be special pleading and unfair because they benefit those with whom one is in status competition. In the latter case, this reaction is likely to take the form of a moral appeal to the state to act as a disciplinary power, to impose an economizing reason upon those who are further down on the social order."[6] Combined with neoliberal economic impulses, might this argument apply as well to immigrants to North America, where the huddled masses of the world's poor are no longer so welcome as they supposedly once were? It apparently applies widely to the conditionalities imposed on Southern states by the IMF and the World Bank.[7] It certainly applies to the insecurities mobilized by advertising agencies to sell all manner of contemporary commodities derived from all parts of the planet. It also applies to the tourism industry where visiting ever more exotic locales, playing a round on high-status golf courses, whether as ecotourists or not, is the status symbol of landscape consumption and the source of a considerable amount of "environmental" knowledge on the part of omnivores who wish states would deal with environmental threats to numerous things.

GLOBALIZING INSECURITY

The globe, so often the symbol of endangerment in environmental thinking, has also been appropriated as a powerful icon by numerous advertisers to peddle all sorts of commodities and invoke a multiplicity of anxieties.[8] Globalization is in this sense a cultural process relating to identities and the symbol of the globe itself, as well as an economic one.[9] In television advertising in North America in the late 1990s, Sprint Canada literally reduced the size of the spinning globe in its commercials that implied that reduced overseas phone call charges made the world a smaller place, and Malibu cars went four times around the world through numerous landscapes before they needed a tune-up. Perhaps most significant for the argument in this book were the Chevy Blazer sport utility vehicles (SUVs) seen negotiating numerous hazards on their journeys through a variety of environments before arriving unscathed either at a beach or at a large suburban home. The vehicles were sold with the slogan "a little security in an insecure world." Security is a matter of a safe space provided technologically to keep external dangers at bay.

But this is an individualist's technological response to a "dangerous" world, a response that, however, only the global corporations can provide. It is one that uses technology to ensure that the autonomous subject and family are not endangered. But there is a very powerful irony here in that the vehicle uses fossil fuels to propel itself and its passengers through storms, the frequency of which may be increased by the global climate changes brought on precisely by the use of fossil fuels. Big vehicles, specifically the popular sport utility vehicles like the Chevy Blazer, are fuel inefficient and, if buying trends at the turn of the century are maintained, will ensure that many states have little hope of meeting carbon dioxide emissions levels agreed to in international climate change agreements in the 1990s. Kaplan's metaphor of the rich in their limousine might be updated to specify sport utility vehicles because the potholes are much worse.

This theme further links to the discussions of neoliberalism, to the contemporary reduction of state functions, and to the provision of public transport. Suburban families in the North increasingly use vehicles to transport children to schools and to numerous other social events. The rationale for using vehicles is frequently an assumption of safety in a political culture of personal insecurity where either strangers are viewed as dangerous predators or vehicles themselves are understood as a threat to pedestrians.[10] In environmental terms, this raises the question of whether the modern subject is sustainable, but it also points very directly to the assumptions about what is to be secured and how security is to be provided. In particular, it suggests once again that the suburban consumer lifestyle aspirations of modern autonomous citizens are premised on an unsustainable global economy. Clearly security in these terms does not take the environment or the insights of ecology very seriously in its specifications of the sources of danger. This concern is especially germane when neoliberal formulations emphasize voluntary and private sector initiatives as a method of ensuring compliance with international climate agreements.[11]

There are further revealing ironies in this formulation of security. In actuarial terms most women who die violently are killed by partners or other family members. Domestic violence suggests that the place specified as most safe is in fact most dangerous. The public sphere, constructed in the contemporary political imaginary as the source of threats, is actually less likely to produce fatal violence. This reasoning is compounded when the death toll due to automobile accidents and pollution-induced respiratory failure is considered. Once again the automobile, sold as the provider of "a little security in an insecure world" is a source of very considerable insecurity.[12] This is especially so because, in a richly ironic reprise of the international relations literature on security dilemmas, these larger "secure" vehicles are likely to do more damage to other vehicles and their occupants in a collision and likely to get into more accidents because of driver overconfidence derived from precisely those feelings of security. This emphasizes the point that security is a political act. In terms of the political specifications of danger, the numbers killed in domestic violence and vehicle accidents are apparently very much less important than the cultural codings of spaces and technologies in terms of safety and danger.

The related point is the importance given to technology and the assumptions that technical acumen can provide the answers to many insecurities. At the turn of the century, sport utility vehicles were frequently marketed in terms not only of security but of giving the driver control of potentially dangerous environments. The domination of nature, within a specifically colonial sensibility, is what many of the vehicle advertisements are all about. Lincoln Navigators, Nissan Pathfinders, Landrover Discoverers, and Ford Explorers and Expeditions all transport their drivers into the wilderness and subdue it by a combination of the technological capabilities of engineering and, of course, the driving prowess of the proud owner. In case such readings are disputed, it might also be noted that in one widely aired Nissan commercial we were told that "Nature is more civilized in a Pathfinder." Other vehicles appropriate the names of large carnivores and domesticate their identities in the process of once again reproducing colonial urban identities and understandings of nature as out there to be subdued. The Acura MDX was introduced as "The Lord of the

Jungle (concrete and otherwise)." In Britain, Landrover's Freelander SUV is released into the wild by African park wardens. In early 2000 Ford Explorers were portrayed towing skiers up a steep mountain slope to the accompanying geopolitical slogan of "no boundaries." A little later Ford Escape SUVs were seen driving on the moon with the assurance that "Escape can conquer even the most foreign terrain." Similar themes play out in numerous other advertising scripts of domination and control. Insecurity and its converse, a metaphysics of domination, however, perpetuate all sorts of insecurities.

HOMO IMPERIALUS?

If the large-scale historical patterns of change, migration, displacement, colonization, and resistance sketched out in this book as the geopolitical framework for understanding the environmental dimensions of contemporary global politics are an accurate reflection of the geohistorical trajectory of humanity, the unavoidable question becomes: are we innately a colonial or imperial species? The answer from many thinkers critical of the scarcity assumption and the blindness of neo-Malthusianism to the precise political economies of misery on the planet is no, not least because of the wise refusal of any claims that there is a single universal human nature. Thus the liberal consumer in search of ever greater commodity-produced "security" is not assumed to be the human telos. But who else we might become is as yet not clear, and the hegemonic aspirations of modernity suggest that the scarcity/insecurity nexus has yet to run its course.[13]

But if the question of imperiality is answered positively, then our apparent need to colonize faces a fundamental limit, the capacity of the biosphere to be reengineered for human life. It is precisely this very limitation that is the focus of the global environmental change literature and most strands of environmentalism. If we are modern, and if modernity is inevitably an urban colonizing system, something has to give. But even many of the counterarguments to "growth," not least as noted in the introduction, such as those from the Worldwatch Institute, remain trapped in economic arguments of scarcity rather than rephrasing the current difficulties in a larger cultural and political framework. Thus in Arturo Escobar's terms:

> Although ecologists and ecodevelopmentalists recognize environmental limits to production, a large number do not perceive the cultural character of the commercialization of nature and life integral to the

Western economy, nor do they seriously account for the cultural limits which many societies have set on unchecked production. It is not surprising that their policies are restricted to promoting the "rational" management of resources. Environmentalists who accept this presupposition also accept imperatives for capital accumulation, material growth, and the disciplining of labor and nature. In doing so they extrapolate the occidental economic culture to the entire universe.[14]

This extrapolation implies that the basic environmental assumption of these discourses, that there is only one Earth, is in fact wrong. This apparently extraordinary assumption is the premise for a remarkable book on the exploration and colonization of Mars by Robert Zubrin. *The Case for Mars* takes the argument about modernity as a colonial project involving ever greater appropriation of nature for human purposes to its logical conclusion.[15] Although the argument that humanity should take to the skies is not new, Zubrin's compilations of ingenious, cheap, and practical suggestions for how to take the first crucial leap to Mars using existing technology are noteworthy, not only because of their sheer inventiveness, but because he directly argues that Mars exploration is analogous to the history of European settlement in the "new world."

One of the results of colonization was a labor shortage that stimulated ingenuity and technological innovation in the United States as well as trade and innovation in the old world of what became known as Europe. Similar possibilities, he argues, exist in the exploration of Mars—both on Mars when settlers get there and back on Earth where research into all sorts of materials and energy sources will be directly stimulated. Even the analogy with the triangle trade from Europe to Africa and thence to the Americas following the winds and currents fits loosely. Technological exports to Mars will facilitate food and rocket fuel production there and provide a supply base for the exploration and mining of concentrated metal ores in the asteroids. Gravity works to the advantage of such a new triangle trade, as getting material from there to Earth is relatively easy. The analogy with the trade winds and westerlies blowing across the Atlantic is apt. Getting supplies from Mars is loosely analogous to using the Caribbean as a supply area for North America, and North America as a source of raw materials for Europe. To this day, at least if the relative sizes of the displays in the local liquor stores are any indication, the descendants of the Scots settlers in Cape Breton drink more rum (from the Caribbean) than scotch (from Scotland)! Although Zubrin doesn't emphasize the point, getting mineral supplies to Earth from the asteroid belt might even reduce the environmental impacts of terrestrial mining.

Once on Mars, it would be necessary to change and thicken the atmosphere to warm the planet to facilitate food production. "Terraforming" is the normal term for discussions of the remaking of astronomical entities to make them suitable for settlement. Designing bacteria to facilitate either gas production or polar icecap albedo changes is one possibility. The use of space mirrors, made of ultrathin reflective fabric and placed over the poles to warm them and speed up the melting of the carbon dioxide ice, is another possible technology. All of this would facilitate changing temperatures and boosting the possibilities for food production while rendering life on the surface easier for settlers. It is interesting that thinking about the possible ecosphere of Mars led James Lovelock to formulate the Gaia hypothesis; and Carl Sagan started his thinking about planetary atmospheres, which led to discussions of nuclear winter and the most compelling links in the 1980s between military action and potential environmental damage, by reflecting on the possibilities of changing the atmosphere on Venus.[16] Such shifts in geopolitical perspective are part of what Zubrin suggests is an important part of innovation. The "discovery" of the Americas radically challenged the cosmology of "Europeans." In a similar way, he argues, a consideration of space travel will change contemporary cosmologies beyond the tropes of a single planetary home.

In turn, thinking in these terms allows us to recognize that the industrial colonization of Earth is, albeit unintentionally, terraforming Earth. "Producing" nature now becomes a much less strange conceptualization of contemporary human processes within the biosphere. Drastic losses of biodiversity, atmospheric change, and so on are forcing natural systems in ways that would not occur without the anthropogenic contribution. The viability of the planet as a home for the species may be compromised by some unanticipated runaway feedback loop, or at least rendered more difficult by disruptions, as compensating natural forces reestablish a dynamic equilibrium in some new state.

The point is not that the planet is a completely stable entity, which it obviously is not, but rather that inadvertent terraforming is what is effectively being done without a plan or any clear idea that this is what is happening. The atmospheric composition of the last ten millennia has been conducive to human civilization; the assumption in environmental thinking, from footprint analysis to global climate change modeling, is that drastically changing it runs counter to the precautionary principle that argues for maintaining this rough homeostatic state. Again, our quest for "security" in modern economic production is currently undermining the conditions for terrestrial habitability. This is the (environmental) security dilemma in its largest form.

But for Zubrin, such pessimism is misreading the history of evolution and of humanity. From the evolution of aerobic bacteria that reduced the carbon dioxide in the terrestrial atmosphere and prevented Earth becoming another Venus—while simultaneously supplying the oxygen that built an ozone layer and providing the conditions for aerobes to survive—through the life forms that colonized the land and formed soil, life has transformed the planet and itself. In an oversimplification concerning ecological productivity, he suggests that humanity follows on from earlier forms of life transforming their environments and so providing the potential for yet further growth.

> Humans are only the most recent practitioners of this art. Starting with our earliest civilizations, we used irrigation, crop seedings, weeding, domestication of animals, and protection of their herds to enhance the activity of those parts of the Earth most efficient in supporting human life. In doing so, we have expanded the biospheric basis for human population, which has expanded our numbers and thereby our power to change our environment to support a continued cycle of exponential growth. As a result, we have literally remade the Earth into a place that can support billions of people, a substantial fraction of whom have been sufficiently liberated from the need to toil for daily survival—such that some can now look out into the night sky for new worlds to conquer.[17]

Given this optimism, not surprisingly Zubrin does not deal with the pressing environmental problems on Earth nor with the huge inequities that allow him to dream of Mars while many others of his species starve. But such considerations do form the backdrop for Kim Stanley Robinson's science fiction trilogy on the colonization of Mars.[18] The point in these novels, which present a detailed working out of a logic similar to that in *The Case for Mars*, is that the terrestrial crisis will form the political and economic backdrop of any colonization, but regardless of how successful it may be, the scale of colonization and the transportation costs of interplanetary travel will ensure that Mars, or other nonterrestrial settlements, cannot provide a large enough safety valve to export terrestrial problems however much technological innovation is stimulated. Mackinder's assumption of closed space still applies.

Mars colonization cannot directly relieve ever-growing populations and economic disparities on Earth, although future technological possibilities opened up by the process will not be predictable. That being the case, disparities will have to be dealt with here on Earth. If the processes of modernization are unstoppable, then the question becomes one of security in the terms that Buzan, Wæver, and de Wilde formulate: environmental security is a matter of the conditions of survival of the necessary ecological conditions for civilization itself.[19] But if civilization is in fact transforming the atmosphere and inducing one of the largest recorded collapses of biodiversity in Earth's geological record, then the process of terraforming that modernity has undertaken is a very dangerous gamble indeed.

The term "terraforming" is apt for the planetary reengineering that we are undertaking. It is unplanned, of course, with no clear indication that the bits of construction will fit together coherently or that the whole project will operate in a way that will not undercut the conditions for habitable life, and the record of large-scale engineering projects in the twentieth century is not reassuring.[20] In many cases side effects or unanticipated consequences produced socially dysfunctional results. Megaprojects operate on a reductionist logic that evades or denies the flexible complexity of natural and social systems. Using such central planning as a mode of dealing with global crisis is unlikely to produce sustainable outcomes, although this is not likely to slow down the aspirations to global management implicit in many of the discussions of global politics in the face of environmental disruptions.[21]

Viewing matters in this way explicitly raises questions about whether the focus on environment ought to be about preserving, restoring, or making landscapes, and whether the new thinking in ecology renders environment redundant.[22] The scale of human impact on ecological systems suggests that we are terraforming inadvertently. Plans to go to Mars and convert its atmosphere into something producing a climate that is suitable for modified terrestrial organisms make this point unavoidable. Preservation is obviously necessary in many places, not least as an attempt to maintain planetary biodiversity. But understanding environment in the negative sense of a residual category or a substrate to humanity, as in the long-standing division of culture and nature, may both misconstrue the important political questions raised by sustainability and perpetuate precisely the colonial geopolitical assumptions that support neo-Malthusian interpretations of political economy.

It is not hard to argue that those who wish to decipher the human genetic code, or patent life forms derived from "bioprospecting" in the gene pools of tropical forests, have already decided that making nature is the way of the future. In this sense colonization continues on the small scale; the very stuff of life is enclosed, commodified, and sold to make a corporate profit.[23] Environmentalist concerns about this have invoked various arguments about the sanctity of nature and life, falling back on the naturalistic language of the culture-nature divide to construct arguments against the corporate appropriation of genetic material. The irony here is that the corporate specification of nature, including human genetic material as intellectual property, and the indigenous assumptions of living in nature implicitly share the same rejection of nature as external to humanity. Charles Darwin long ago pointed out that the human-nature divide was ontologically untenable. But the politics of who decides what shape life will take in this "postmodern" condition becomes only more pressing in the context of abandoning nature as the supposed external refuge for principles of law and morality.

This returns the argument to discussions of sustainability, what should be sustained, and who should decide on these matters. If humanity is inadvertently terraforming the planet, or *Remaking Reality*, as the title of a recent collection of essays on the topic has it, then the politics of security become more directly concerned with the ontological categories of our being.[24] Globalization makes the interconnections of different places unavoidable in constructing a consideration of politics. The assumptions of autonomous spaces are no more convincing than the narrative of progress as development. No longer can the geopolitical sleight of hand, where space is turned into time, their difference represented as a matter of their being more primitive on an evolutionary path than modern us, work to occlude connections and responsibilities.[25]

For Zubrin, innovation and colonization is precisely what needs to be maintained: "I would say that failure to terraform Mars constitutes failure to live up to our human nature and a betrayal of our responsibility as members of the community of life itself."[26] This can be reinterpreted slightly in considering the colonization of Mars in metaphoric terms as either the process of (inadvertent?) parthogenesis, where Gaia divides to grow, or as the pessimistic assumption that disaster awaits us in the metastasis of the human cancer tumor on the earthly substrate.[27] Which metaphor you prefer depends on the assumptions made as to the place of humanity in the great order of things, a philosophical and ontological question that is increasingly unavoidable on a small endangered planet. Because precisely what is being secured is the crucial, if frequently unasked, question underlying the discussions of broadening security. What structures nearly all these discussions about the subject of security is the neoliberal autonomous consumer.

The human security agenda draws its list of political desiderata from a similar series of assumptions and usually fails to note that the various antidotes to threats to security in its formulations are not necessarily additive. Responding to economic insecurities within the existing system may threaten community or health security. The roots of both the more narrowly economistic formulations of liberalism in terms of neoliberalism and the more broadly inclusive liberalism underlying the broadened security agenda emphasize the autonomous subject without fully considering the ecological questions of their existence. Reimagining who we might become also requires coming to terms with the terms within which we construct our own political and cultural identities. In particular, the question of ecological situatedness requires thinking through the questions of colonization.

COLONIZATION

The history of colonization has been one of resistance, frequently articulated in the tropes of justice when not in terms of claims to cultural autonomy and the rights to self-determination. One of the problems with the technological colonization discourse is that it short-circuits discussions of justice and equity and assumes that further colonization is the human telos. To an ecologist attuned to the consequences of the transformation of diverse ecosystems into agricultural monocultures, Zubrin's technological optimism may read more like hubris than a description of natural evolution. Nonetheless, considered in light of the discussion in chapters 5 and 7, as a narrative of "progress," his specification of humanity as just the latest colonial lifeform to drastically change its environment is a very useful benchmark against which to compare other formulations of the human predicament.

The civil society arguments at the Rio Earth Summit and subsequently have emphasized the rampant injustices in environmental matters and the often drastic consequences of what Baechler summarizes in chapter 3 as maldevelopment.[28] To address environmental security in terms of the prevention of the destruction of livelihoods and lives among the most marginal environmental refugees is to discuss the agenda of postdevelopment and focus on the casualties of development. But if security is living in a condition of freedom from threats, then the processes of colonization seem destined to continue to endanger marginal peoples as they become omnivores, whether wishing in some sense to do so or not. The assumption that the planet can be converted into a resource-supplying entity for the growing urban population, all of whom aspire to contemporary carboniferous lifestyles, is what footprint analysis and discussions of environmental space render untenable.

This circle has to be squared as Thomas Homer-Dixon emphasizes. To do so requires ingenuity and adaptability. But does it have to come from a reinvention of colonization in Zubrin's terms, or are there other possible forms of ingenuity available? In Walker's terms from the epigraph to this chapter, is it possible to become other than what we have already become? Does modernity inevitably mean that the rest of the world has to follow American models of the good life, a life of fossil-fueled material consumption? Can ingenuity be indigenous knowledge and technical options that are precisely the forms of understanding excluded from consideration by the assumption that scientific technical knowledge is the only form of thinking that is likely to provide solutions? But Zubrin also understands that the successful explorer lives off the land and learns the local ways to survive and colonize. Ecological sensitivity is about carefully learning the specifics of context; traditional knowledge is about precisely these insights.

The Wuppertal program suggests that alternatives are possible once the interconnections and the distant consequences of local actions are understood. The history of economic growth through the use of carboniferous fuel has been one of increased efficiencies at least as far as energy consumption per GNP unit is concerned.[29] But as numerous ecological economists, in addition to the Wuppertal Institute authors, have repeatedly pointed out, more drastic institutional changes are needed, especially in North America, to encourage the adoption of more efficient technologies, not to mention less-damaging land use patterns and reduced-impact transportation systems.[30] The Wuppertal suggestions, as with many other environmental projects, operate in ways that think about ecological connections, not in ways that necessarily require either universal agreement on programs or complex institutional coordination.

But scholars and activists need a vocabulary that deals with the human condition on the largest scale too. As this book has suggested, neither the conventional understandings of environment nor the traditional conceptual tools of international relations specifically and the social sciences more generally offer many helpful suggestions. If the colonizing nature of humanity is not a given, despite Zubrin's insightful, if oversimplified, "whig" summary of ecological history, and there is no determined future to our fate, then the crucial questions of how we rethink politics and the purpose of our collective endeavors sensitive to the ecological geopolitics require further thought. The question for security studies, in particular, is whether there is, in Homer-Dixon's apt phrasing from chapter 3, a large enough supply of ingenuity in the postmodern academy to rethink the assumptions of security. To conflate the themes in the larger work of the authors of the first and last epigraphs to this book, can we rescue it from metaphysics, certainty, and force and reimagine security as openness, change, connection, and becoming? In other words, can the conceptualizations move from physics to ecology, geopolitics to ecopolitics, being on Earth to living within a biosphere that the rich and powerful among us are rapidly changing?

Using Gadgil and Guha's vocabulary of ecosystem people, omnivores, and environmental refugees has the advantage of clearly positioning people in terms of their ecological situation. It has the further advantage of relative simplicity, producing a typology that is flexible and applicable to numerous situations. The obligation of the rich omnivores to recognize that their consumption has distant impacts on the ecosystem people is clarified by considerations of footprints and shadows. As this volume has tried to suggest, it is the rich omnivores whose actions have the greatest impacts within the biosphere and whose cultural codes of rights to consumption so powerfully reinforce the destruction. Not all ecosystem people wish to remain so. Urbanization is a disruptive force, but one that also caters to the aspirations of numerous people who wish to "be other than they have already become."[31]

For all the discussion of globalization, postmodernism, diasporas, and postcolonialism, perhaps the overarching change in the last century was one driven by geographical relocation to the cities and into what it is now possible to suggest, as Magnusson does, is analogous to a single global city. The assumptions that this can continue to be literally driven by fossil-fueled vehicles is no longer tenable once the discussions of contemporary research into climate change are examined closely.[32] But the human condition is increasingly an urban one, and the politics of what is so inadequately understood in terms of environment has to take this much more seriously, not least because the traditions of politics that so shape our thinking are premised on the urban assumptions of Athens and Rome even as they continue to conflate claims of autonomous community with the rights implicit in empire.

GEOPOLITICS

Specifically, if we are living in a time of an emerging global city, the assumptions that the most important geopolitical categories are national frontiers may be fundamentally wrong. If this is a time of flows and electronic communication, perhaps the maps that matter most are those of the interconnection patterns between urban nodes. Manuel Castells has helpfully suggested a network society to encompass many dimensions of this pattern.[33] Humanity is migrating from the countryside to the towns in the process of becoming omnivores, and then moving up the urban hierarchy from smaller centers to the major nodes of the global economy. Some of the omnivores are leaving the biggest cities to telecommute from "rural" areas whose ecologies they further change as they consume its landscapes. As Connolly and Kennedy's analysis, discussed briefly in chapter 5, suggests, the border guards and immigration officials who regulate transnational flows are often simply dealing with the effects of borders that happen to be in the way of these movements.

Coupled to a shadow analysis of the flows of resources in the South from rural areas to urban areas and ports, a pattern of global resource and population flows emerges that is only partly, and poorly, caught in the assumptions of states as the most important categories. Viewed in terms of the global city, the resource peripheries are part of the infrastructure of the metropolis; center and periphery are interlinked in terms of the raw materials, agricultural commodities, and environmental services the periphery provides. This is the logic of imperialism caught in the contemporary, popular renditions of these things at the beginning of the twenty-first century, from the structure of Microsoft's computer game *Age of Empires,* through episodes of the *Star Wars* movie series, to the colonizing premises of Zubrin's *The Case for Mars.* To extend the metaphors of earlier chapters it might also be understood as a system where the Mobutos of this world act as the Indian agents of southern parts of the imperial system.

Similar thinking suggests to John Agnew that mapping power in the post–cold war world is not effectively done by thinking in terms of states.[34] It might in many ways be better done in terms of connecting nodes rather than reproducing boundaries. States are then much less relevant to understanding the key processes that occur, however important their claims to legal jurisdiction or military necessity may be in managing the local arrangements that administer the larger patterns of flows. If such thinking is coupled to Johan Galtung's ideas of the structure of imperialism, where the elites of North and South have common connected interests that allow them to effectively play the poor of their respective states against each other, this gives us a rough political model of power that emphasizes the international mobility of the rich omnivorous cosmopolitans arranging long-distance resource flows to maintain their affluence and political control.[35] Such thinking suggests parameters for discussing environmental problems with reference to a more adequate geopolitical imaginary than that presented by assumptions of relatively autonomous spatial units.

Perhaps the democratically benign peace theory is actually about active integration in the world economy, as Etel Solignen suggests.[36] Since many of the states at the core of the world trading system are also democratic in at least the formal sense, is the democratic peace actually the imperial peace of the rich and powerful? The findings by Baechler, discussed in chapter 3, that environmental conflicts are related to groups that do not have any other choice, as is the case for many ecosystem people, supports this contention. It doesn't follow that all people who have alternative economic options are not forced to fight, but clearly desperation is a powerful factor in the choice to use violence. States with complex industrial infrastructures and liberal heritages that resist the use of force are obviously often deterred from using force against similar states, but the temptations to use violence against technologically inferior military forces remains an option, especially now that these are effectively bloodless events for the industrial powers using cruise missiles and air power.[37]

In Mary Kaldor's terms, the globalization of warfare suggests that these matters are interconnected.[38] She argues that the new wars are about smaller collectivities mobilized around identity politics in the face of collapsing states and global economic disruptions. These wars tend to destroy existing state structures and economies. They are financed by the fund-raising efforts of diasporic populations at a distance, coupled to arms trading and the wholesale looting of environments. They are unlikely to happen in the zones of peace of the affluent, but are all too common in the peripheral areas of the global economy. But, as pointed out in chapter 5, the resources stripped and exported to pay for wars and buy arms mostly produced in the affluent North are, in at least a few prominent examples, part of the processes of environmental degradation that concern those thinking about the intersections of environment and conflict. They are also accelerating the processes set in motion by the colonizing practices supporting urban development.

ECOPOLITICS

Focusing on ecological endangerment in a global urban system suggests repeatedly that the geopolitical precepts of security thinking and the implicit assumptions in conventional environmental thinking severely limit the formulations of political alternatives and the identities that might be secured in the future. The possibilities of autonomous, self-sufficient polities and the invocation of local community have frequently been used in environmental discourses, as in so many other expressions of contemporary ontotheological aspiration, and undoubtedly have considerable utility in practical struggles in many rural areas. But to uncritically repeat these Aristotelian formulations of the polis as political ideal, given the scale of the global ecological footprint and the pace of carboniferous modernization, suggests both a failure of geopolitical imagination and a nostalgia for a nonexistent past that cannot offer a model for secure lives in the future in many urbanizing places. Moral spaces, to invoke the title of a recent compilation of thinking on this theme, have to include a recognition of the ethical relations between people at a distance, omnivores and ecosystem people, if they are to encapsulate modern identities in ways that are sensitive to interconnections and in ways that extend also to nonhuman entities, as in the aboriginal formulations of these things in terms of "all our relations."[39]

The Wuppertal Institute's focus on the connections between North and South, or more specifically between particular places and institutions, recognizes the advantage in thinking in local terms because of the reduction in fuel expenditures and in the disruptions due to transportation infrastructure as well as in using ecologically sensitive production systems, but it refuses the nostalgia of the local because the importance of distant interconnections is now undeniable in contemporary urban settings. The political utility of this is obvious insofar as it renders geopolitical invocations of security threats more difficult. Mobilizing "we" against "them" is made at least a little more difficult by discourses and practical economic and institutional arrangements that focus on the consequences of ecological interconnections. The crucial point is that environmental modernization, or a discussion of environment in terms only of pollution and "end of the pipe" technical fixes, fails to tackle the scale of resource throughput, or the extractions of specific distant resources, and is an inadequate formulation of "environment" for thinking about ecology and security in any sense.

This is especially important if the response to ecological change produces either a militaristic response, as suggested in Allen Hammond's recent scenario of a future "fortress world," or a violent response in Malthusian terms against immigrants and the urban poor, as Mike Davis suggests is a logical extrapolation of the contemporary fears and moral panics among the affluent populations driving sport utility vehicles and living in the gated communities of Southern California.[40] Repeated attempts to corral the dump bears and the environmental refugees in a carceral landscape of technological surveillance suggest one possible pessimistic model of the future. Given the increasing trends to private security provision, such considerations are important in how ecological crisis is portrayed.

Rethinking political identities to emphasize the interconnectedness of people in distant places is not easy. It probably has to involve cultural shifts from conspicuous consumption to what Peter Taylor calls conspicuous aestheticism, where what is desirable, as well as praiseworthy, is ecological efficiency and a minimization of material throughputs in an economy.[41] Quite how such a cultural change is to be accomplished remains a major political question for anyone with green political inclinations and is especially fraught in the neoliberal climate of manufactured anxiety and risk society. Perhaps a little ironically, the detailed plans for the extension of human colonization into the solar system, and in particular the practical arrangements for keeping humans alive on Mars, emphasize thinking very carefully about ecological interconnections. They do so by drawing attention to the geopolitical presupposition that there is only one Earth. While this is an appropriate assumption in present circumstances, there is no reason to suppose that it will hold for centuries to come.

Planning a project such as the colonization of Mars also draws attention back to the important point made in chapter 4 about the appropriate timescale for thinking about environments. If global warming is driven by the use of fossil fuels that have effectively been contributing to climate change for two hundred years, then it is reasonable to think about the implications of contemporary decisions about these matters in terms of a couple of centuries. This is the period that North American aboriginal thinking emphasizes in its formulation of matters in terms of contemporary decisions being judged in light of the likely consequences for the seventh generation hence.[42] While predictability is forever a problem, the seventh-generation criterion is a useful template for considering the consequences of ecological policies precisely because it forces attention on the long-term possibilities and what options current actions may foreclose for future generations. But whatever preferred futures many people might specify, it is clear that the liberal, supposedly autonomous individual, enmeshed in a fossil-fueled economy of unconstrained consumption, is not tenable into the far future.[43]

What the postcolonial and postmodern subject of security might aspire to in its various urban locations is unspecifiable as such; indeed, the very possibility of such an emergence would seem to be far from the considerations of many of the currently rich and powerful and their advertising agencies busy celebrating the technologically capable consumer as the extension of liberal autonomy and simultaneously the subject of multiple insecurities. But the assumption of autonomy is increasingly untenable in a small biosphere. The illusion that we are on Earth, rather than part of the biosphere, is increasingly challenged by scientific discourse as well as practical wisdom. The social movements opposing global financial and trade organizations are repeatedly pointing to the unavoidable contradictions of assuming that autonomous states and individuals constitute the only possible political vocabulary.

Innovation and ingenuity may be key to future adaptations in many places, but the important point from the analysis in this book is that both are tied to the identity of postcolonial subjects—not just postcolonial in the sense of belonging to societies trying to reinvent themselves in the aftermath of formal political independence from European imperialism, but societies that are also trying to reimagine how to live sensitive to the politics and ecology within a biosphere recently seriously endangered by its imperial residents.[44] Thus this is also a matter of postdevelopment. Zubrin may be right; the future may lie with those who can imagine a future of space travel and ever-expanding human horizons. Space may be the final frontier, one that allows infinite expansion of human power, one that facilitates such things as an end in themselves.[45] But in that case the poor will probably remain expendable; the supposed necessity of getting to the stars may well require the destruction of this planet and the appropriation of others less well suited to human habitation.

The largest questions of humanity's future cannot avoid the matter of who decides on what is most valuable. This is the big question that haunts the environmental security discourse but that is so frequently deferred in technical discussions and avoided in calls for short-term policy relevance for the would-be global managers. Is what is most valuable a viable planetary biosphere for a civilization (or civilizations) that lives well within the ecological possibilities of a small planet? Or is the planet to be used as a resource base to construct a technological, but very definitely colonial, future with a continuation or acceleration of all the inequities and political violence that this has so far engendered? Presumably the future will not be either of these visions winning a straightforward global political struggle between the ecologists and the colonizers. But phrasing it in these terms has the advantage of focusing attention squarely on the questions of identity, on who we are as a result of the interconnections that enmesh our lives, and of what kind of world we want to hand on to subsequent generations.

How this question is answered depends to a very large extent on whether that "we" refers to a rich omnivore, a marginalized eco-system person, or a desperate environmental refugee. So far the discussion of such matters in terms of environmental security has been written by the rich omnivores in their comfortable offices and libraries. The occupants of Kaplan's limousine are the theorists of insecurity. It has not been, at least until very recently, written by the aboriginal peoples turned into environmental refugees. It has not been written by the residents of the wild zones outside the white picket fences of suburban bliss. But there is no guarantee that if they were given the tools to write scholarly texts on environmental security, they might not also buy into the colonizing gamble.

Security for whom is now very much a matter of where the insecurity lies. The critical turn in security studies makes this geopolitical question unavoidable. But the extension of these matters to the broader agenda of post–cold war security studies now makes clear that the assumptions of emancipation implicit in most of the critical literature need a more explicit engagement with the ecological conditions of contemporary urban existence. To do so requires much further hard thinking about the liberal identities supposedly endangered by various "environmental" threats, as well as about the colonial vocabularies within which geopolitical thought in general, and security discourse in particular, remain enmeshed.

Notes

INTRODUCTION

1. See Michael Klare and Yogesh Chandrani, eds., *World Security: Challenges for a New Century* (New York: St. Martin's Press, 1998).

2. Felix Dodds, ed., *Earth Summit 2002: A New Deal* (London: Earthscan, 2000); Michael Klare, *Resource Wars: The New Landscape of Global Conflict* (New York: Metropolitan Books, 2001).

3. K. Krause and Michael C. Williams, "Broadening the Agenda of Security Studies: Politics and Methods," *Mershon International Studies Review* 40, no. 2 (1996): 229–54; Barry Buzan, Ole Wæver, and Jaap de Wilde, *Security: A New Framework for Analysis* (Boulder, Colo.: Lynne Rienner, 1998).

4. D. B. Bobrow, "Complex Insecurity: Implications of a Sobering Metaphor," *International Studies Quarterly* 40 (1996): 435–50.

5. P. B. Stares, ed., *The New Security: A Global Survey* (Tokyo: Japan Center for International Exchange, 1998).

6. Michael Dillon, *Politics of Security: Towards a Political Philosophy of Continental Thought* (London: Routledge, 1996); see also Mark Neocleous, "Against Security," *Radical Philosophy* 100 (2000): 7–14.

7. Ole Wæver, "Figures of International Thought: Introducing Persons instead of Paradigms," in Iver B. Newmann and Ole Wæver, eds., *The Future of International Relations: Masters in the Making?* (London: Routledge, 1997).

8. Keith Krause and Michael C. Williams, "Preface: Toward Critical Security Studies," in Krause and Williams, eds., *Critical Security Studies: Concepts and Cases* (Minneapolis: University of Minnesota Press, 1997), ix.

9. Ken Booth, "Security and Emancipation," *Review of International Studies* 17, no. 4 (1991): 313–26; Richard Wyn-Jones, *Security, Strategy, and Critical Theory* (Boulder, Colo.: Lynne Rienner, 1999).

10. Yosef Lapid and Friedrich Kratochwil, eds., *The Return of Culture and Identity in IR Theory* (Boulder, Colo.: Lynne Reinner, 1995).

11. Peter J. Katzenstein, ed., *The Culture of National Security: Norms and Identity in World Politics* (New York: Columbia University Press, 1996).

12. J. Weldes, M. Laffey, H. Gusterson, and R. Duvall, eds., *Cultures of Insecurity: States, Communities, and the Production of Danger* (Minneapolis: University of Minnesota Press, 1999).

13. Cynthia Enloe, *Maneuvers* (Berkeley: University of California Press, 1999); Carol Cohn, "Sex and Death in the Rational World of Defence Intellectuals," *Signs: Journal of Women in Culture and Society* 12 (1987): 687–718; and M. Cooke and A. Woollacott, eds., *Gendering War Talk* (Princeton, N.J.: Princeton University Press, 1993).

14. Karen Litfin, ed., *The Greening of Sovereignty in World Politics* (Cambridge: MIT Press, 1998).

15. R. B. J. Walker, *Inside/Outside: International Relations as Political Theory* (Cambridge: Cambridge University Press, 1993).

16. Gearóid Ó Tuathail and Simon Dalby, *Rethinking Geopolitics* (London: Routledge, 1998); Ó Tuathail, Simon Dalby, and Paul Routledge, eds., *The Geopolitics Reader* (London: Routledge, 1998); Mark Polelle, *Raising Cartographic Consciousness: The Social and Foreign Policy Vision of Geopolitics in the Twentieth Century* (Lanham, Md.: Lexington Books, 1999); Klaus Dodds and David Atkinson, *Geopolitical Traditions: A Century of Geopolitical Thought* (London: Routledge, 2000).

17. On geographs, see Simon Dalby, "The 'Kiwi Disease': Geopolitical Discourse in Aotearoa/New Zealand and the South Pacific," *Political Geography* 12, no. 5 (1993): 437–56. For similar studies using formulations of spatial metaphors, see James Sidaway, "Geopolitics, Geography, and 'Terrorism' in the Middle East," *Environment and Planning D: Society and Space* 12, no. 3 (1994): 357–72; and on the spatial narratives of foreign policy, see Klaus Dodds, "War Stories: British Elite Narratives of the 1982 Falklands/Malvinas War," *Environment and Planning D: Society and Space* 11 (1993): 619–40.

18. Gearóid Ó Tuathail and John Agnew, "Geopolitics and Discourse: Practical Geopolitical Reasoning in American Foreign Policy," *Political Geography* 11 (1992): 190–204.

19. Simon Dalby, *Creating the Second Cold War: The Discourse of Politics* (New York: Guilford, 1990).

20. Gearóid Ó Tuathail, *Critical Geopolitics: The Politics of Writing Global Space* (Minneapolis: University of Minnesota Press, 1996).

21. John Agnew, *Geopolitics: Revisioning World Politics* (London: Routledge, 1998).

22. Geoffrey Parker, *Geopolitics: Past, Present, and Future* (London: Pinter, 1998).

23. Klaus Dodds, *Geopolitics in Antarctica: Views from the Southern Oceanic Rim* (New York: Wiley, 1997).

24. This is not to argue that geography is irrelevant, but rather that ontological specifications based on geopolitical premises are important in political terms. Geography in the sense of topographical physiography is unavoidably a matter of strategy, as Colin Gray yet again emphasizes in a recent essay. But the question of geopolitical reasoning in security discussions and claims to endangerment focuses on the political implications of the prior geopolitical specification of identities rather than the consequences of such formulations for matters of military operations and miscalculations. See Colin S. Gray, "Inescapable Geography," in Gray and Geoffrey Sloan, eds., *Geopolitics: Geography and Strategy* (London: Frank Cass, 1999), 161–77. However, for a critique of geopolitics that directly challenges Gray's earlier formulations of American cold war nuclear strategy, see Dalby, *Creating the Second Cold War.*

25. David Campbell, *Writing Security: United States Foreign Policy and the Politics of Identity*, 2d ed. (Minneapolis: University of Minnesota Press, 1998).

26. Martin W. Lewis and Karen E. Wigen, *The Myth of Continents: A Critique of Metageography* (Berkeley: University of California Press, 1997), ix.

27. Joanne Sharp, *Condensing the Cold War: Reader's Digest and American Identity* (Minneapolis: University of Minnesota Press, 2000).

28. David Campbell, *National Deconstruction: Violence, Identity, and Justice in Bosnia* (Minneapolis: University of Minnesota Press, 1998).

29. Daniel Deudney, "Ground Identity: Nature, Place, and Space in Earth Nationalism," in Lapid and Kratochwil, *The Return of Culture and Identity in IR Theory*, 129–45.

30. Simon Dalby, "Geopolitics and Global Security: Culture, Identity, and the 'Pogo Syndrome,'" in Ó Tuathail and Dalby, *Rethinking Geopolitics*, 295–313.

31. Neil Smith, *Uneven Development: Nature, Capital, and the Production of Space* (Oxford: Blackwell, 1990).

32. Michael Shapiro, *Violent Cartographies: Mapping Cultures of War* (Minneapolis: University of Minnesota Press, 1997).

33. I sketched a similar argument in "Reading Rio, Writing the World: *The New York Times* and the Earth Summit," *Political Geography* 15, no. 6/7 (1996): 593–614.

34. This has subsequently been reprinted as the title essay in a collection

of Kaplan's political writings: *The Coming Anarchy: Shattering the Dreams of the Post Cold War* (New York: Random House, 2000).

35. See Immanuel Wallerstein, "Ecology and Capitalist Costs of Production," in his *The End of the World As We Know It: Social Science for the Twenty-First Century* (Minneapolis: University of Minnesota Press, 1999), 76–86; Alf Hornborg, "Towards an Ecological Theory of Unequal Exchange: Articulating World System Theory and Ecological Economics," *Ecological Economics* 25 (1998): 127–36; Alf Hornborg, "Ecosystems and World Systems: Accumulation as an Ecological Process," *Journal of World Systems Research* 4 (1998): 169–77.

36. See the overviews of these literatures in Susanne Jakobsen, "International Relations and Global Environmental Change," *Cooperation and Conflict* 34, no. 2 (1999): 205–36; and Matthew Paterson, "Interpreting Trends in Environmental Global Governance," *International Affairs* 75, no. 4 (1999): 793–802.

37. See Ronnie Lipschutz, *After Authority: War, Peace, and Global Politics in the Twenty-First Century* (Albany: State University of New York Press, 2000).

38. Richard Falk, *Predatory Globalization: A Critique* (Cambridge, England: Polity Press, 1999).

39. Michael Hardt and Antonio Negri, *Empire* (Cambridge: Harvard University Press, 2000).

40. See Alex Wendt, *Social Theory of International Relations* (Cambridge: Cambridge University Press, 1999); "Forum on Social Theory and International Relations," *Review of International Studies* 26, no. 1 (2000): 123–80; Jenny Edkins, *Poststructuralism and International Relations: Bringing the Political Back In* (Boulder, Colo.: Lynne Rienner, 1999); and Nalini Persram, "Coda: Sovereignty, Subjectivity, Strategy," in Jenny Edkins, Nalini Persram, and Veronique Pin-Fat, eds., *Sovereignty and Subjectivity* (Boulder, Colo.: Lynne Rienner, 1999), 163–75.

41. Michel Foucault, "Practicing Criticism," in *Politics, Philosophy, Culture: Interviews and Other Writings, 1977–1984* (New York: Routledge, 1988), 154–55; Campbell, *Writing Security*, 191.

42. Campbell, *Writing Security*; Ronnie Lipschutz, ed., *On Security* (New York: Columbia University Press, 1995); Paul Chilton, *Security Metaphors: Cold War Discourse from Containment to Common House* (New York: Peter Lang, 1996); Karen Fierke, *Changing Games, Changing Strategies* (Manchester: Manchester University Press, 1998).

43. Eric Laferrière and Peter J. Stoett, *International Relations Theory and Ecological Thought* (London: Routledge, 1999); Matthew Paterson, *Understanding Global Environmental Politics* (London: Macmillan, 2000); Jon Barnett, *The Meaning of Environmental Security: Ecological Politics and Policy in the New Security Era* (London: Zed, 2001).

44. See, for instance, Robert L. Heilbroner, *An Inquiry into the Human Prospect* (New York: Norton, 1974); William Ophuls, *Ecology and the Politics of Scarcity: Prologue to a Political Theory of the Steady State* (San Francisco: Freeman, 1977); Richard J. Barnet, *The Lean Years: Politics in the Age of Scarcity* (New York: Simon and Schuster, 1980); and the critique in Francis Sandbach, *Environment, Ideology, and Policy* (Oxford: Blackwell, 1980).

45. Neil Evernden, *The Social Creation of Nature* (Baltimore: Johns Hopkins University Press, 1992). See, in general, Wolfgang Sachs, *Planet Dialectics: Explorations in Environment and Development* (London: Zed, 1999).

46. The politics of characterizing the "environment" are now widely discussed elsewhere: see, for instance, M. Jimmie Killingsworth and Jacqueline S. Palmer, *Ecospeak: Rhetoric and Environmental Politics in America* (Carbondale: Southern Illinois University Press, 1992); Jane Bennett and William Chaloupka, eds., *In the Nature of Things: Language, Politics, and the Environment* (Minneapolis: University of Minnesota Press, 1993); Karen Litfin, *Ozone Discourses: Science and Politics in Global Environmental Cooperation* (New York: Columbia University Press, 1994); John S. Dryzek, *The Politics of the Earth: Environmental Discourses* (Oxford: Oxford University Press, 1997); Eric Darrier, ed., *Discourses of the Environment* (Oxford: Blackwell, 1999).

47. Whatever their ideological preoccupations, such publications as the annual reports from the Worldwatch Institute or the biannual summary from the World Resources Institute are sobering reading in terms of anthropogenic actions within a dynamic biosphere. See Lester Brown et al., *State of the World* (New York: Norton, annual); World Resources Institute, *World Resources 1998–1999* (Oxford: Oxford University Press, 1998). The now classic analyses in B. L. Turner et al., *The Earth as Transformed by Human Action* (Cambridge: Cambridge University Press, 1990), can also be complemented with such publications as United Nations Environment Programme, *Global Environmental Outlook 1997* (Oxford: Oxford University Press, 1997), to suggest some of the more important dimensions of the contemporary crisis.

48. Timothy W. Luke, *Ecocritique: Contesting the Politics of Nature, Economy, and Culture* (Minneapolis: University of Minnesota Press, 1997); specifically on media coverage of these matters and the corporate specifications of the terms for debate, see Moti Nissani, "Media Coverage of the Greenhouse Effect," *Population and Environment: A Journal of Interdisciplinary Studies* 21, no. 1 (1999): 27–43.

49. Peter J. Taylor, *Modernities: A Geohistorical Interpretation* (Minneapolis: University of Minnesota Press, 1999); Jonathan Crush, ed., *Power of Development* (London: Routledge, 1995); Arturo Escobar, *Encountering Development* (Princeton, N.J.: Princeton University Press, 1995).

50. Roxanne Lynn Doty, *Imperial Encounters: The Politics of Representation in North-South Relations* (Minneapolis: University of Minnesota Press, 1996).

51. Jeremy Rifkin, *Biospheric Politics: A New Consciousness for a New Century* (New York: Crown, 1991); Partha Chatterjee and Matthias Finger, *The Earth Brokers: Power, Politics, and World Development* (London: Routledge, 1994); Maarten A. Hajer, *The Politics of Environmental Discourse: Ecological Modernization and the Policy Process* (Oxford: Oxford University Press 1995); Alan Russell and John Vogler, eds., *The International Politics of Biotechnology: Investigating Global Futures* (Manchester: Manchester University Press, 2000).

52. Luke, *Ecocritique*, 85.

53. Braden R. Allenby, "Environmental Security: Concept and Implementation," *International Political Science Review* 21, no. 1 (2000): 5–21.

1. RETHINKING SECURITY STUDIES

1. Samuel Huntington, "The Clash of Civilizations," *Foreign Affairs* 72, no. 3 (1993): 22–49; Samuel Huntington, *The Clash of Civilizations and the Remaking of World Order* (New York: Simon and Schuster, 1996).

2. Zbigniew Brzezinski, *The Grand Chessboard: American Primacy and Its Geostrategic Imperatives* (New York: Basic Books, 1997).

3. Benjamin Barber, *Jihad vs. McWorld: How Globalism and Tribalism Are Reshaping the World* (New York: Ballantine, 1996).

4. A. Herod, G. Ó Tuathail, and S. M. Roberts, eds., *Unruly World? Globalization, Governance, and Geography* (London: Routledge, 1998).

5. Richard J. Barnet and J. Cavanagh, *Global Dreams: Imperial Corporations and the New World Order* (New York: Simon and Schuster, 1994).

6. Kenichi Ohmae, *The End of the Nation State: The Rise of Regional Economies* (New York: Free Press, 1995).

7. John Agnew and Stuart Corbridge, *Mastering Space: Hegemony, Territory, and International Political Economy* (London: Routledge, 1995).

8. Etel Solingen, *Regional Orders at Century's Dawn: Global and Domestic Influences on Grand Strategy* (Princeton, N.J.: Princeton University Press, 1998).

9. Mary Kaldor, *New and Old Wars: Organized Violence in a Global Era* (Stanford, Calif.: Stanford University Press, 1999).

10. J. J. Romm, *Defining National Security: The Nonmilitary Aspects* (New York: Council on Foreign Relations, 1993).

11. James Rosenau, *Along the Domestic-Foreign Frontier: Exploring Governance in a Turbulent World* (Cambridge: Cambridge University Press, 1997).

12. Yosef Lapid and Friederich Kratochwil, eds., *The Return of Culture and Identity in IR Theory* (Boulder, Colo.: Lynne Reinner, 1995); Keith Krause and Michael C. Williams, eds., *Critical Security Studies: Concepts and Cases* (Minneapolis: University of Minnesota Press, 1997).

13. See the *Economist* survey, "The New Geopolitics," 31 July 1999.

14. Max Cameron, Robert Lawson, and Brian Tomlin, eds., *To Walk without Fear: The Global Movement to Ban Landmines* (Toronto: Oxford University Press, 1998).

15. R. B. J. Walker, "The Subject of Security," in Keith Krause and Michael Williams, *Critical Security Studies: Concepts and Cases* (Minneapolis: University of Minnesota Press, 1997), 62.

16. Simon Dalby, "Contesting an Essential Concept: Reading the Dilemmas in Contemporary Security Discourse," in Krause and Williams, *Critical Security Studies,* 3–31.

17. Bradley Klein, *Strategic Studies and World Order: The Global Politics of Deterrence* (Cambridge: Cambridge University Press, 1994).

18. Ken Booth, ed., *Statecraft and Security: The Cold War and Beyond* (Cambridge: Cambridge University Press, 1998).

19. Michael MccGwire, *Perestroika and Soviet National Security* (Washington, D.C.: Brookings Institution, 1991).

20. Saul H. Mendlovitz and R. B. J. Walker, eds., *Towards a Just World Peace: Perspectives from Social Movements* (London: Butterworth, 1987).

21. J. Ann Tickner, *Gender in International Relations: Feminist Perspectives on Achieving Global Security* (New York: Columbia University Press, 1992); Cynthia Enloe, *Bananas, Beaches, and Bases: Making Feminist Sense of International Politics* (London: Pandora, 1989).

22. Richard Ullman, "Redefining Security," *International Security* 8, no. 1 (1983): 133.

23. D. B. Bobrow, "Complex Insecurity: Implications of a Sobering Metaphor," *International Studies Quarterly* 40 (1996): 435–50.

24. Michael Klare and Yogesh Chandrani, eds., *World Security: Challenges for a New Century* (New York: St. Martin's Press, 1998).

25. Seyom Brown, "World Interests and the Changing Dimensions of Security," in Klare and Chandrani, *World Security,* 1–17.

26. Barry Buzan, Ole Wæver, and Jaap de Wilde, *Security: A New Framework for Analysis* (Boulder, Colo.: Lynne Rienner, 1998). See also Barry Buzan, *People, States, and Fear: An Agenda for International Security Studies in the Post–Cold War Era* (Boulder, Colo.: Lynne Rienner, 1991).

27. Buzan et al., *Security,* 7.

28. Caroline Thomas and Peter Wilkin, eds., *Globalization, Human Security, and the African Experience* (Boulder, Colo.: Lynne Rienner, 1999); Peter Stoett, *Human and Global Security: An Exploration of Terms* (Toronto: University of Toronto Press, 1999).

29. United Nations Development Program, *Human Development Report 1994* (New York: Oxford University Press, 1994), 24.

30. Ibid., 23.

31. Ibid., 34.

32. See, for instance, the Commission on Global Governance, *Our Global Neighbourhood* (Oxford: Oxford University Press, 1995); and the Independent Commission on Population and Quality of Life, *Caring for the Future* (Oxford: Oxford University Press, 1996). In terms of the environment, Michael Renner invokes the UNDP framework in his book *Fighting for Survival: Environmental Decline, Social Conflict, and the New Age of Insecurity* (New York: Norton, 1996).

33. Emma Rothschild, "What Is Security?" *Daedalus* 124, no. 3 (1995): 55.

34. See the discussion in K. Krause and Michael C. Williams, "Broadening the Agenda of Security Studies: Politics and Methods," *Mershon International Studies Review* 40, no. 2 (1996): 229–54.

35. David Baldwin, "The Concept of Security," *Review of International Studies* 23, no. 1 (1997): 5–26.

36. Steven Walt, "The Renaissance of Security Studies," *International Studies Quarterly* 35 (1991): 211–39.

37. Ole Wæver, "Securitization and Desecuritization," in R. Lipschutz, ed., *On Security* (New York: Columbia University Press, 1995), 46–86.

38. Carl Schmitt, *The Concept of the Political* (Chicago: University of Chicago Press, 1996). Originally published as *Der Begriff des Politischen*, 1932.

39. See, in particular, R. B. J. Walker, *Inside/Outside: International Relations as Political Theory* (Cambridge: Cambridge University Press, 1993).

40. R. B. J. Walker, "International Relations and the Fate of the Political," in Michi Ebata and Beverly Neufeld, eds., *Confronting the Political in International Relations* (London: Macmillan, 2000), 212–38.

41. Baldwin, "Concept"; he cites Arnold Wolfers, "'National Security' as an Ambiguous Symbol," *Political Science Quarterly* 67 (1952), reprinted in Arnold Wolfers, "National Security as an Ambiguous Symbol," in *Discord and Collaboration: Essays in International Politics* (Baltimore: Johns Hopkins University Press, 1962).

42. Jef Huysmans, "Security! What Do You Mean? From Concept to Thick Signifier," *European Journal of International Relations* 4, no. 2 (1998): 226–55.

43. Ibid., 232.

44. Michael Dillon, *Politics of Security: Towards a Political Philosophy of Continental Thought* (London: Routledge, 1996).

45. Michael C. Williams, "Identity and the Politics of Security," *European Journal of International Relations* 4, no. 2 (1998): 205.

46. Ibid., 213.

47. Rothschild, "What Is Security?"

48. James Richardson, "Contending Liberalisms: Past and Present," *European Journal of International Relations* 3, no. 1 (1997): 5–33; David Long, *Towards a New Liberal Internationalism: The International Theory of J. A. Hobson* (Cambridge: Cambridge University Press, 1996).

49. Robert Latham, *The Liberal Moment: Modernity, Security, and the Making of Postwar International Order* (New York: Columbia University Press, 1997).

50. Ken Booth, "Security and Emancipation," *Review of International Studies* 17, no. 4 (1991): 313–26; Richard Wyn-Jones, *Security, Strategy, and Critical Theory* (Boulder, Colo.: Lynne Rienner, 1999).

51. Richard Ashley, "Sovereignty, Hauntology, and the Mirror of the World Political: Some Thoughts Too Long Re-Tained," paper presented at the annual convention of the International Studies Association, Los Angeles, March 2000.

52. David Campbell, *Writing Security: United States Foreign Policy and the Politics of Identity,* 2d ed. (Minneapolis: University of Minnesota Press, 1998); Benedict Anderson, *Imagined Communities: Reflections on the Origin and Spread of Nationalism* (London: Verso, 1991).

53. Campbell, *Writing Security,* 170.

54. P. B. Stares, ed., *The New Security: A Global Survey* (Tokyo: Japan Center for International Exchange, 1998).

55. Jon Barnett, "Destabilizing the Environment-Conflict Thesis," *Review of International Studies* 26, no. 2 (2000): 271–88.

56. Klein, *Strategic Studies and World Order.*

57. Mohammed Ayoob, *The Third World Security Predicament: State Making, Regional Conflict, and the International System* (Boulder, Colo.: Lynne Rienner, 1995).

58. Keith Krause, "Insecurity and State Formation in the Global Military Order: The Middle Eastern Case," *European Journal of International Relations* 2, no. 3 (1996): 319–54.

59. Mustapha Pasha, "Security as Hegemony," *Alternatives* 21, no. 3 (1996): 287. See also, in detail on South Asia, Sankaran Krishna, *Postcolonial Insecurities: India, Sri Lanka, and the Question of Nationhood* (Minneapolis: University of Minnesota Press, 1999).

60. Larry A. Swatuk and Peter Vale, "Why Democracy Is Not Enough: Southern Africa and Human Security in the Twenty-first Century," *Alternatives* 24, no. 3 (1999): 361–89.

61. See Dan Smith, *The State of War and Peace Atlas* (London: Penguin, 1997).

62. Carnegie Commission on Preventing Deadly Conflict, *Preventing Deadly Conflict: Final Report* (New York: Carnegie Corporation, 1997), 17.

63. Gunther Baechler, *Violence through Environmental Discrimination: Causes, Rwanda Arena, and Conflict Model* (Dordrecht: Kluwer, 1999). See also K. Volden and D. Smith, eds., *Causes of Conflict in the Third World* (Oslo: International Peace Research Institute, 1997).

64. As will be elaborated in chapter 2, concerns about resources and nature and the political dimensions of these things are of long-standing concern and not only to the followers of Thomas Malthus. The themes of environmental security were discussed in the 1970s and 1980s, before it was so labeled. See Richard Falk, *This Endangered Planet: Prospects and Proposals for Human Survival* (New York: Random House, 1971). Lester Brown, *Redefining National Security*, Worldwatch Paper 14 (Washington, D.C.: Worldwatch Institute, 1977), is probably the most direct precursor to the late 1980s discussion; see also Dennis Pirages, *The New Context for International Relations: Global Ecopolitics* (North Scituate, Mass.: Duxbury Press, 1978); W. H. Durham, *Scarcity and Survival in Central America: Ecological Origins of the Soccer War* (Stanford, Calif.: Stanford University Press, 1979); and Lloyd Timberlake and Jon Tinker, "The Environmental Origins of Political Conflict," *Socialist Review* 15, no. 6 (1985): 57–75. The theme was important in discussions of the World Commission on Environment and Development in the 1980s and is incorporated in the commission's final report, *Our Common Future* (Oxford: Oxford University Press, 1987).

65. Francis Fukuyama, "The End of History," *National Interest* 16 (1989): 3–18.

66. Norman Myers, "Environment and Security," *Foreign Policy* 47 (1989): 23–41; Jessica T. Mathews, "Redefining Security," *Foreign Affairs* 68, no. 2 (1989): 162–77; Michael Renner, *National Security: The Economic and Environmental Dimensions*, Worldwatch Paper 89 (Washington, D.C.: Worldwatch Institute, 1989).

67. Neville Brown, "Climate, Ecology and International Security" *Survival* 31, no. 6 (1989): 519–32; Peter Gleick, "The Implications of Global Climatic Changes for International Security," *Climatic Change* 15 (1989): 309–25; Arthur H. Westing, "The Environmental Component of Comprehensive Security," *Bulletin of Peace Proposals* 20, no. 2 (1989): 129–34; Josh Karliner, "Central America's Other War," *World Policy Journal* 6 (1989): 787–810.

68. Patricia Mische, "Ecological Security and the Need to Reconceptualise Sovereignty," *Alternatives* 14, no. 4 (1989): 389–427; on the Brazil discussion specifically, see Jose Goldemberg and Eunice Ribeiro Durham, "Amazonia and National Sovereignty," *International Environmental Affairs* 2, no. 1 (1990): 22–39.

69. Gwyn Prins and R. Stamp, *Top Guns and Toxic Whales: The Environment and Global Security* (London: Earthscan, 1991). Other popular works included Jeremy Rifkin, *Biospheric Politics: A New Consciousness for a New*

Century (New York: Crown, 1991); and later Norman Myers, *Ultimate Security: The Environmental Basis of Political Stability* (New York: Norton, 1993).

70. Daniel Deudney, "The Case against Linking Environmental Degradation and National Security," *Millennium* 19 (1990): 461–76; reprinted in slightly updated form as "The Mirage of Ecowar: The Weak Relationship among Global Environmental Change, National Security, and Interstate Violence," in I. H. Rowlands and M. Greene, eds., *Global Environmental Change and International Relations* (London: Macmillan, 1992). These themes are trenchantly restated in Daniel Deudney, "Environmental Security: A Critique," in Daniel Deudney and Richard Matthew, eds., *Contested Grounds: Security and Conflict in the New Environmental Politics* (Albany: State University of New York Press, 1999), 187–219.

71. Peter H. Gleick, "Environment and Security: The Clear Connections," *Bulletin of the Atomic Scientists* 47, no. 3 (1991): 16–21; Daniel Deudney, "Environment and Security: Muddled Thinking," *Bulletin of the Atomic Scientists* 47, no. 3 (1991): 22–28.

72. Patricia Mische, "Security through Defending the Environment: Citizens Say Yes," in E. Boulding, ed., *New Agendas for Peace Research: Conflict and Security Reexamined* (Boulder, Colo.: Lynne Rienner, 1992), 103–19; Lothar Brock, "Security through Defending the Environment: An Illusion?" in ibid., 79–102.

73. Matthias Finger, "The Military, the Nation State, and the Environment," *Ecologist* 21, no. 5 (1991): 220–25.

74. See the discussions collected in Jyrki Kakonen, ed., *Perspectives on Environmental Conflict and International Relations* (London: Pinter 1992); Gwyn Prins, ed., *Threats without Enemies: Facing Environmental Insecurity* (London: Earthscan, 1993); and Jyrki Kakonen, ed., *Green Security or Militarized Environment* (Aldershot: Dartmouth, 1994). More recent reiterations of these debates, with some of the same contributing authors, are in Deudney and Matthew, eds., *Contested Grounds,* and M. Lowi and B. Shaw, eds., *Environment and Security: Discourses and Practices* (London: Macmillan, 2000).

75. See, for instance, M. Tennberg, "Risky Business: Defining the Concept of Environmental Security," *Cooperation and Conflict* 30, no. 3 (1995): 239–58; K. Dokken and N. Graeger, *The Concept of Environmental Security: Political Slogan or Analytical Tool?* (Oslo: Peace Research Institute, Oslo, 1995); Geoffrey D. Dabelko and David Dabelko, "Environmental Security: Issues of Conflict and Redefinitions," *Environment and Security* 1, no. 1 (1996): 23–49; Robyn Eckersley, "Environmental Security Dilemmas," *Environmental Politics* 5 (1996): 140–46; Lorraine Elliot, "Environmental Conflict: Reviewing the Arguments," *Journal of Environment and*

Development 5, no. 2 (1996): 149–67; Nina Graeger, "Environmental Security," *Journal of Peace Research* 33 (1996): 109–16. Geoffrey D. Dabelko and P. J. Simmons, "Environment and Security: Core Ideas and US Government Initiatives," *SAIS Review* 17, no. 1 (1997): 127–46; G. B. Thomas, "U.S. Environmental Security Policy: Broad Concern or Narrow Interests," *Journal of Environment and Development* 6, no. 4 (1997): 397–425; Lothar Brock, "The Environment and Security: Conceptual and Theoretical Issues," in N. P. Gleditsch, ed., *Conflict and the Environment* (Dordrecht: Kluwer, 1997), 17–34; N. P. Gleditsch, "Armed Conflict and the Environment: A Critique of the Literature," *Journal of Peace Research* 35, no. 3 (1998): 381–400; Betsy Hartmann, "Population, Environment, and Security: A New Trinity," in J. M. Silliman and Y. King, eds., *Dangerous Intersections: Feminist Perspectives on Population, Environment, and Development* (Boston: South End 1999), 1–23; Ruth E. Noorduyn and Wouter T. deGroot, "Environment and Security: Improving the Interaction of Two Science Fields," *Journal of Environment and Development* 8, no. 1 (1999): 24–48; Karen T. Litfin, "Constructing Environmental Security and Ecological Interdependence," *Global Governance* 5, no. 3 (1999): 359–77; Richard Matthew, "The Environment as a National Security Issue," *Journal of Policy History* 12, no. 1 (2000): 101–22; Jeroen Warner, "Global Environmental Security: An Emerging 'Concept of Control,'" in Philip Stott and Sian Sullivan, eds., *Political Ecology: Science, Myth and Power* (London: Arnold, 2000), 247–65.

76. For a detailed discussion of the concept of environmental security and much more on these nuances, see Jon Barnett, *The Meaning of Environmental Security: Ecological Politics and Policy in the New Security Era* (London: Zed, 2001).

77. Marc Levy, "Is the Environment a National Security Issue?" *International Security* 20 (1995): 35–62.

78. Thomas Homer-Dixon, "On the Threshold: Environmental Changes as Causes of Acute Conflict," *International Security* 16, no. 1 (1991): 76–116.

2. THE ENVIRONMENT AS GEOPOLITICAL THREAT

1. Robert D. Kaplan, "The Coming Anarchy," *Atlantic Monthly* 273, no. 2 (1994): 44–76.

2. Halford J. Mackinder, "The Geographical Pivot of History," *Geographical Journal* 23, no. 4 (1904), reprinted in R. Kasperson and J. Minghi, eds., *The Structure of Political Geography* (Chicago: Aldine, 1969).

3. It apparently received a wide readership. The article was reproduced

in the *San Francisco Chronicle* on 13 March 1994, and was commented on by media columnists including Anthony Lewis in the *New York Times* ("A Bleak Vision," 7 March 1994, A17). It has been cited, in a diverse range of languages, in articles concerned in one way or another with visions of the future, in academic and policy journals ranging from sociological theory in the Czech *Sociologicky Casopis* to design philosophy in *Ergonomics*. Random House publishers reissued it as the title article in a book six years after it first appeared: Kaplan, *The Coming Anarchy: Shattering the Dreams of the Post Cold War* (New York: Random House, 2000). In the preface to the book version, Kaplan says that the article was "translated into over a dozen languages and reprinted constantly" (xv).

4. G. Meyers, T. Klak, and T. Koehl, "The Inscription of Difference: News Coverage of the Conflicts in Rwanda and Bosnia," *Political Geography* 15, no. 1 (1996): 21–46; Cyril Obi, "Globalized Images of Environmental Security in Africa," *Review of African Political Economy* 83 (2000): 47–62.

5. I. Bellany, "Malthus and the Modern World," *Review of International Studies* 20, no. 4 (1994): 411–22. The crucial text is Thomas Malthus, *An Essay on the Principle of Population* (1798; reprint, Harmondsworth, England: Penguin, 1970).

6. Simon Dalby, "The Threat from the South? Geopolitics, Equity, and Environmental Security," in Daniel Deudney and Richard Matthew, eds., *Contested Grounds: Security and Conflict in the New Environmental Politics* (Albany: State University of New York Press, 1999), 155–85; Vandana Shiva, "Conflicts of Global Ecology: Environmental Activism in a Period of Global Reach," *Alternatives* 19, no. 2 (1994): 195–207.

7. R. D. Kaplan, *Balkan Ghosts* (New York: St. Martin's Press, 1993). Elizabeth Drew suggests that *Balkan Ghosts* had a considerable effect on President Clinton's Bosnia policy in *On the Edge: The Clinton Presidency* (New York: Simon and Schuster, 1994).

8. J. Rosner, "Is Chaos America's Real Enemy?" *Washington Post,* 14 August 1994, C1–C2.

9. Marc Levy, "Is the Environment a National Security Issue?" *International Security* 20 (1995): 35; President Clinton's remarks to the National Academy of Sciences, 29 June 1994, are in the Woodrow Wilson Center, *Environmental Change and Security Project Report* 1 (1995): 51.

10. Geoffrey D. Dabelko and P. J. Simmons, "Environment and Security: Core Ideas and US Government Initiatives," *SAIS Review* 17, no. 1 (1997): 136.

11. Michel Foucault, "Governmentality," in G. Burchell, C. Gordon, and P. Miller, eds., *The Foucault Effect* (Chicago: Chicago University Press, 1991), 87–104.

12. Richard Grove, *Ecology, Climate, and Empire: Colonialism and Global Environmental History, 1400–1940* (Cambridge, England: White Horse, 1997).

13. Eric B. Ross, *The Malthus Factor: Poverty, Politics, and Population in Capitalist Development* (London: Zed, 1998).

14. Nicholas Xenos, *Scarcity and Modernity* (London: Routledge, 1989).

15. Harrison Brown, *The Challenge of Man's Future* (New York: Viking, 1954). There is an interesting precursor to Kaplan's use of Fukuyama (see below) in the bibliographic essay at the end of Brown's book. He also cites a title with the term "posthistoric" in it: R. Seidenberg, *Posthistoric Man, an Inquiry* (Chapel Hill: University of North Carolina Press, 1950).

16. Paul R. Ehrlich, *The Population Bomb* (New York: Ballantine, 1968).

17. D. H. Meadows, D. L. Meadows, J. Randers, and W. W. Behrens III, *The Limits to Growth* (London: Pan, 1974).

18. David Harvey, "Population, Resources, and the Ideology of Science," *Economic Geography* 50, no. 3 (1974): 256–77. On China, Harvey cites W. Vogt, *The Road to Survival* (New York: Sloan, 1948).

19. Paul R. Ehrlich and Anne H. Ehrlich, *The Population Explosion* (New York: Simon and Schuster, 1990).

20. D. H. Meadows, D. L. Meadows, and J. Randers, *Beyond the Limits* (London: Earthscan, 1992).

21. Vaclav Smil, "How Many People Can the Earth Feed?" *Population and Development Review* 20, no. 2 (1994): 255–92.

22. J. R. Wilmoth and P. Ball, "The Population Debate in American Popular Magazines," *Population and Development Review* 18, no. 4 (1992): 631–68.

23. Gearóid Ó Tuathail and T. W. Luke, "Present at the (Dis)integration: Deterritorialization and Reterritorialization in the New Wor(l)d Order," *Annals of the Association of American Geographers* 84, no. 3 (1994): 381–98. They note that "wild" and "tame" zones can be read from Samuel Huntington's "The Clash of Civilizations" (*Foreign Affairs* 72, no. 3 [1993]: 22–49); and are a particularly salient theme in M. Singer and A. Wildavsky, *The Real World Order: Zones of Peace, Zones of Turmoil* (Chatham, N.J.: Chatham House, 1993).

24. On environmental determinism and the "Soviet threat," see Simon Dalby, *Creating the Second Cold War: The Discourse of Politics* (New York: Guilford, 1990); in general, see Ronnie Lipschutz, "Environmental Conflict and Environmental Determinism: The Relative Importance of Social and Natural Factors," in N. P. Gleditsch, ed., *Conflict and the Environment* (Dordrecht: Kluwer, 1997), 35–50.

25. Michael Shapiro, *The Politics of Representation* (Madison: University

of Wisconsin Press, 1988), 100; Michael Shapiro, *Reading the Postmodern Polity* (Minneapolis: University of Minnesota Press, 1992).

26. There is a vast literature on the themes of the domination of nature and its philosphical roots in works such as William Leiss, *The Domination of Nature* (Boston: Beacon, 1974), and Carolyn Merchant, *The Death of Nature: Women, Ecology, and the Scientific Revolution* (San Francisco: Harper and Row, 1980), as well as such themes as pastoral ideals in American thought in Leo Marx, *The Machine in the Garden* (New York: Oxford University Press, 1964). Undoubtedly, Kaplan's analysis in part finds its resonance with its audience because of the cultural ambiguities in the multiple constructions of "nature" and "wilderness."

27. Kaplan, "The Coming Anarchy," 58.

28. Ibid., 48.

29. Ibid., 52.

30. See Laurie Garrett, *The Coming Plague: Newly Emerging Diseases in a World out of Balance* (New York: Farrar Strauss, 1994).

31. Kaplan, "The Coming Anarchy," 54.

32. See, in general, Norman Myers, *Ultimate Security: The Environmental Basis of Political Stability* (New York: Norton, 1993); Dennis Pirages, "Demographic Change and Ecological Security," in Michael T. Klare and Daniel C. Thomas, eds., *World Security: Challenges for a New Century* (New York: St. Martin's Press, 1994), 314–31.

33. Thomas Homer-Dixon, "On the Threshold: Environmental Changes as Causes of Acute Conflict," *International Security* 16, no. 1 (1991): 76–116; George Kennan (alias "X"), "The Sources of Soviet Conduct," *Foreign Affairs* 25, no. 4 (1947): 566–82.

34. Kaplan, "The Coming Anarchy," 59.

35. Kaplan's article went to press before the appearance of Homer-Dixon's later *International Security* article that modified his earlier hypotheses (see Thomas Homer-Dixon, "Environmental Scarcities and Violent Conflict: Evidence from Cases," *International Security* 19, no. 1 [1994]: 5–40); but after the more popular version of these findings was published in Thomas Homer-Dixon, J. H. Boutwell, and G. W. Rathjens, "Environmental Change and Violent Conflict," *Scientific American* 268, no. 2 (1993): 38–45. See also Thomas Homer-Dixon, "Environmental and Demographic Threats to Canadian Security," *Canadian Foreign Policy* 2, no. 2 (1994): 7–40. These later articles show that as Homer-Dixon's research progressed, he further qualified the earlier tentative hypotheses on which Kaplan builds his alarmist Malthusian interpretation. Kaplan's summary is much more obviously Malthusian than Homer-Dixon's published research.

36. Kaplan, "The Coming Anarchy," 60.

37. Samuel Huntington, "The Clash of Civilizations," *Foreign Affairs* 72, no. 3 (1993): 22–49.

38. Gearóid Ó Tuathail, "Putting Mackinder in His Place: Material Transformations and Myth," *Political Geography* 11, no. 1 (1992): 100–118.

39. Martin van Creveld, *The Transformation of War* (New York: Free Press, 1991).

40. See N. Gibbs, "Why? The Killing Fields of Rwanda," *Time*, 16 May 1994, 21–27. In a reflection of Kaplan's themes, the caption under the heading of this feature article reads, "Hundreds of thousands have died or fled in a month of tribal strife. Are these the wars of the future?"

41. Gearóid Ó Tuathail, *Critical Geopolitics: The Politics of Writing Global Space* (Minneapolis: University of Minnesota Press, 1996).

42. Kaplan, "The Coming Anarchy," 59.

43. Ibid., 45.

44. Fantu Cheru, "Structural Adjustment, Primary Resource Trade, and Sustainable Development in Sub-Saharan Africa," *World Development* 20, no. 4 (1992): 497–512.

45. J. Warnock, *The Politics of Hunger: The Global Food System* (London: Methuen, 1987).

46. Fiona Mackenzie, "Exploring the Connections: Structural Adjustment, Gender, and the Environment," *Geoforum* 24, no. 1 (1993): 71–87.

47. On the complicated interconnections of these themes in West Africa in the early 1990s, see T. Shaw and J. E. Okolo, eds., *The Political Economy of Foreign Policy in ECOWAS* (New York: St. Martin's Press, 1994).

48. This "amnesia" is a recurring feature in many development discourses; see David Slater, "The Geopolitical Imagination and the Enframing of Development Theory," *Transactions of the Institute of British Geographers*, n.s., 18 (1993): 419–37; Jonathan Crush, ed., *Power of Development* (London: Routledge, 1995).

49. R. L. Paarlberg, "The Politics of Agricultural Resource Abuse," *Environment* 36, no. 8 (1994): 7–9, 33–42. Some of the relations of modernization and rural conflict in Africa were explored in Olivia Bennet, ed., *Greenwar: Environment and Conflict* (London: Panos, 1991), well before Kaplan penned his article.

50. Homer-Dixon's ideas are discussed in chapter 3; in general, see also D. J. Hogan, "The Impact of Population Growth on the Physical Environment," *European Journal of Population* 8 (1992): 109–23.

51. Kaplan, "The Coming Anarchy," 46. Islam is "Western" at least insofar as it is monotheistic in comparision to the animism and complex pantheons of some African beliefs.

52. R. D. Kaplan, "Into the Bloody New World: A Moral Pragmatism for

America in an Age of Mini-Holocausts," *Washington Post*, 17 April 1994, C1–C2.

53. S. Cohen, "Global Geopolitical Change in the Post–Cold War Era," *Annals of the Association of American Geographers* 81, no. 4 (1991): 551–80; S. Cohen, "Geopolitics in the New World Era: A New Perspective to an Old Discipline," in G. J. Demko and W. B. Wood, eds., *Reordering the World: Geopolitical Perspectives on the Twenty-first Century* (Boulder, Colo.: Westview, 1994), 15–48.

54. See, for instance, A. Adadeji, ed., *Africa within the World: Beyond Dispossession and Dependance* (London: Zed, 1993); D. R. F. Taylor and F. Mackenzie, eds., *Development from Within: Survival in Rural Africa* (London: Routledge, 1992); and more generally, Samir Amin, *Maldevelopment: Anatomy of a Global Failure* (London: Zed, 1990).

55. John Agnew, "The Territorial Trap: The Geographical Assumptions of International Relations Theory," *Review of International Political Economy* 1, no. 1 (1994): 53–80.

56. See, for instance, E. Anderson, "The Geopolitics of Military Material Supply," *GeoJournal* 31, no. 2 (1993): 207–13; and D. Volman, "Africa and the New World Order," *Journal of Modern African Studies* 31, no. 1 (1993): 1–30. The counterargument was that resource access issues were of declining importance due to technological innovation and global markets; see R. D. Lipschutz, *When Nations Clash: Raw Materials, Ideology, and Foreign Policy* (Cambridge, Mass.: Ballinger, 1989).

57. N. L. Whitehead and R. B. Ferguson, eds., *War in the Tribal Zone: Expanding States and Indigenous Warfare* (Santa Fe, N.M.: School of American Research, distributed by University of Washington Press, 1992).

3. ENVIRONMENT, CONFLICT, AND VIOLENCE

1. World Commission on Environment and Development, *Our Common Future* (Oxford: Oxford University Press, 1987), 290.

2. Daniel Deudney, "The Mirage of Ecowar: The Weak Relationship among Global Environmental Change, National Security, and Interstate Violence," in I. H. Rowlands and M. Greene, eds., *Global Environmental Change and International Relations* (London: Macmillan, 1992), 169–91; T. Homer-Dixon, M. Levy, G. Porter, and J. Goldstone, "Debate," *Environmental Change and Security Project Report* 2 (1996): 49–71; N. P. Gleditsch, "Armed Conflict and the Environment: A Critique of the Literature," *Journal of Peace Research* 35, no. 3 (1998): 381–400.

3. For a variety of perspectives in a large literature, see Arthur H.

Westing, ed., *Environmental Hazards of War: Releasing Dangerous Forces in an Industrialised World* (London: Sage, 1990); M. Sadiq and J. C. McCain, *The Gulf War Aftermath: An Environmental Tragedy* (Dordrecht: Kluwer, 1993); Matthias Finger, "Global Environmental Degradation and the Military," in Jyrki Kakonen, ed., *Green Security or Militarized Environment* (Aldershot: Dartmouth, 1994), 169–91; Arthur H. Westing, ed., "Armed Forces and the Environment," special issue of *Environment and Security* 1, no. 2 (1997); J. David Singer and Jeffrey Keating, "Military Preparedness, Weapon Systems, and the Biosphere: A Preliminary Impact Statement," *New Political Science* 21, no. 3 (1999): 325–43.

4. Marc Levy, "Is the Environment a National Security Issue?" *International Security* 20 (1995): 35–62.

5. As argued in part in Norman Myers, *Ultimate Security: The Environmental Basis of Political Stability* (New York: Norton, 1993); and Michael Renner, *National Security: The Economic and Environmental Dimensions,* Worldwatch Paper 89 (Washington, D.C.: Worldwatch Institute, 1989).

6. Michael Renner, *Fighting for Survival: Environmental Decline, Social Conflict, and the New Age of Insecurity* (New York: Norton, 1996).

7. Commission on Global Governance, *Our Global Neighbourhood* (Oxford: Oxford University Press, 1995).

8. Deudney, "The Mirage of Ecowar"; Karen Litfin, *Ozone Discourses: Science and Politics in Global Environmental Cooperation* (New York: Columbia University Press, 1994).

9. William Wood, "Forced Migration: Local Conflicts and International Dilemmas," *Annals of the Association of American Geographers* 84, no. 4 (1994): 607–34; Astri Suhrke, "Environmental Degradation, Migration, and the Potential for Violent Conflict," in N. P. Gleditsch, ed., *Conflict and the Environment* (Dordrecht: Kluwer, 1997), 255–72; Steve Lonergan, "The Role of Environmental Degradation in Population Displacement," *Environmental Change and Security Project Report* 4 (1998): 5–15.

10. Thomas Homer-Dixon, "Environmental Scarcities and Violent Conflict: Evidence from Cases," *International Security* 19, no. 1 (1994): 5–40; Homer-Dixon, "Environmental Scarcity, Mass Violence, and the Limits to Ingenuity," *Current History* 95, no. 604 (November 1996): 359–65; Homer-Dixon, "The Project on Environment, Population, and Security: Key Findings of Research," *Environmental Change and Security Project Report* 2 (1996): 45–48.

11. An especially pointed exchange of views on these matters appeared in Homer-Dixon and M. Levy, "Correspondence: Environment and Security," *International Security* 20 (1995/96): 189–98.

12. T. Homer-Dixon, "Strategies for Studying Causation in Complex

Ecological-Political Systems," *Journal of Environment and Development* 5 (1996): 132–48.

13. Levy, "Is the Environment a National Security Issue?"

14. C. F. Ronnfeldt, "Three Generations of Environment and Security Research," *Journal of Peace Research* 34, no. 4 (1997): 473–82.

15. K. M. Lietzmann and G. D. Vest, eds., *Environment and Security in an International Context,* report no. 232 (Bonn: North Atlantic Treaty Organization Committee on the Challenges of Modern Society, 1999).

16. Daniel C. Esty, "Pivotal States and the Environment," in Robert Chase, Emily Hill, and Paul Kennedy, eds., *The Pivotal States: A New Framework for U.S. Policy in the Developing World* (New York: Norton, 1999), 290–314. The original proposal is outlined in Robert Chase, Emily Hill, and Paul Kennedy, "Pivotal States and U.S. Strategy," *Foreign Affairs* 75, no. 1 (1996): 33–51.

17. T. Homer-Dixon, *Environment, Scarcity, and Violence* (Princeton, N.J.: Princeton University Press, 1999), 5.

18. Ibid., 48.

19. Ibid., 73.

20. Ibid., citing J. Leonard, "Overview," in *Environment and the Poor: Development Strategies for a Common Agenda* (New Brunswick, N.J.: Transaction, 1989), 5–7.

21. Homer-Dixon, *Environment, Scarcity, and Violence,* 81.

22. Although an important caveat is obviously needed here, as wars to increase territorial control can fairly easily be said to be about increasing agricultural land. The Nazi plans of lebensraum in Eastern Europe were, after all, in part about gaining agricultural land. However, and this is important for contemporary rather than historical consideration, the increasingly robust international norms that now prohibit wars of territorial aggrandizement, coupled to the point that, at least for states with developed economies, trade and industrial development are more effective strategies for wealth accumulation, suggest that land is no longer a major factor in warfare even if some minor boundary readjustments still occur. On the apparent end of territorial aggrandizement, see Mark W. Zacher, "The Territorial Integrity Norm: International Boundaries and the Use of Force," *International Organization* 55, no. 2 (2001): 215–50.

23. Valerie Percival and T. Homer-Dixon, "Environmental Scarcity and Violent Conflict: The Case of South Africa," *Journal of Peace Research* 35, no. 3 (1998): 279–98.

24. See Ashok Swain, "Displacing the Conflict: Environmental Destruction in Bangladesh and Ethnic Conflict in India," *Journal of Peace Research* 33, no. 2 (1996): 189–204.

25. P. Howard and T. Homer-Dixon, "The Case of Chiapas, Mexico," in T. Homer-Dixon and J. Blitt, eds., *Ecoviolence: Links among Environment, Population, and Security* (Lanham, Md.: Rowman and Littlefield, 1998), 19–65.

26. Homer-Dixon, *Environment, Scarcity, and Violence,* 10.

27. This is the starting point of the "Pivotal States" analysis: Chase, Hill, and Kennedy, *The Pivotal States.*

28. E. Economy, *Environmental Scarcities, State Capacity, Civil Violence: The Case of China* (Cambridge, Mass.: American Academy of Arts and Sciences, 1997); Vaclav Smil and Mao Yushi, eds., *Environmental Scarcities, State Capacity, Civil Violence: The Economic Costs of China's Environmental Degradation* (Cambridge, Mass.: American Academy of Arts and Sciences, 1998).

29. On China's fate, see Jack Goldstone, "The Coming Chinese Collapse," *Foreign Policy* 99 (1995): 35–53; and the rejoinder by Yasheng Huang, "Why China Will Not Collapse," *Foreign Policy* 99 (1995): 54–68. See also Lester R. Brown, *Who Will Feed China? Wake Up Call for a Small Planet* (New York: Norton, 1995); Vaclav Smil, "China's Environment and Security: Simple Myths and Complex Realities," *SAIS Review* 17, no. 1. (1997): 107–26; and Woodrow Wilson Center Environmental Change and Security Project, *China Environment Series.*

30. Alana Boland, "Feeding Fears: Competing Discourses of Interdependency, Sovereignty, and China's Food Security," *Political Geography* 19, no. 1 (2000): 55–76.

31. C. V. Barber, *Environmental Scarcities, State Capacity, Civil Violence: The Case of Indonesia* (Cambridge, Mass.: American Academy of Arts and Sciences, 1997). On the forest fires and their connection to economic and political crisis, see N. Marinova, "Indonesia's Fiery Crises," *Journal of Environment and Development* 8, no. 1 (1999): 70–81.

32. See also T. Homer-Dixon, "The Ingenuity Gap: Can Poor Countries Adapt to Resource Scarcity?" *Population and Development Review* 21, no. 3 (1995): 587–612; and more generally, T. Homer-Dixon, *The Ingenuity Gap* (New York: Knopf, 2000).

33. Homer-Dixon, *Environment, Scarcity, and Violence,* 44.

34. G. A. Baechler and K. R. Spillmann, eds., *Kreigsursache Umweltzerstorung: Environmental Degradation as a Cause of War,* 3 vols. (Zurich: Verlag Ruegger, 1996).

35. Gunther Baechler, *Violence through Environmental Discrimination: Causes, Rwanda Arena, and Conflict Model* (Dordrecht: Kluwer, 1999), 55.

36. Ibid., 61.

37. See also Gunther Baechler, "Why Environmental Transformation Causes Violence: A Synthesis," *Environmental Change and Security Project*

Report 4 (1998): 24–44; Gunther Baechler, "Environmental Degradation and Violent Conflict: Hypotheses, Research Agendas, and Theory Building," in M. Suliman, ed., *Ecology, Politics, and Violent Conflict* (London: Zed, 1999), 76–112; and Gunther Baechler, "Environmental Degradation in the South as a Cause of Armed Conflict," in A. Carius and K. M. Lietzmann, eds., *Environmental Change and Security: A European Perspective* (Berlin: Springer, 1999), 107–29.

38. Baechler, *Violence through Environmental Discrimination,* 85, 86.

39. Ibid., 86, 87.

40. Ibid., 100, 102.

41. Ibid., 99.

42. Ibid., 100. At least in the case of South Africa, Homer-Dixon's analysis concurs with the ENCOP conclusion about violence within an ethnic group; see Percival and Homer-Dixon, "Environmental Scarcity and Violent Conflict: The Case of South Africa."

43. Baechler, *Violence through Environmental Discrimination,* 226.

44. Ibid., 221.

45. The focus on the Sahel is not new; the ENCOP researchers are part of a longer tradition of concern about this region. See Olivia Bennet, ed., *Greenwar: Environment and Conflict* (London: Panos, 1991). But simplistic assumptions of desertification and degradation here as elsewhere in Africa need to be avoided. Recent research shows these environments to be much more complex than popular models suggest. See Melissa Leach and Robin Mearns, eds., *The Lie of the Land: Challenging Received Wisdom on the African Environment* (Oxford: James Currey, 1996).

46. The question of the degradation of mountain environments, which Baechler largely avoids, is an especially fraught issue because Malthusian assumptions of degradation caused by population growth so inaccurately portray ecological and social processes there. See Michael Thompson, "The New World Disorder: Is Environmental Security the Cure?" *Mountain Research and Development* 18, no. 2 (1998): 117–22; and Michael Thompson, "Not Seeing the People for the Population: A Cautionary Tale from the Himalaya," in Miriam Lowi and Brian Shaw, eds., *Environment and Security: Discourses and Practices* (London: Macmillan, 2000), 192–206.

47. Baechler, *Violence through Environmental Discrimination,* 227.

48. Karl Polanyi, *The Great Transformation: The Political and Economic Origins of Our Time* (Boston: Beacon, 1957), 181.

49. On Kenya in particular, see Colin H. Kahl, "Population Growth, Environmental Degradation, and State Sponsored Violence: The Case of Kenya, 1991–93," *International Security* 23, no. 3 (1998): 80–119.

50. Amita Baviskar, *In the Belly of the River: Tribal Conflicts over Development in the Narmada Valley* (Delhi: Oxford University Press, 1995).

51. Bruce Rich, *Mortgaging the Earth: The World Bank, Environmental Impoverishment, and the Crisis of Development* (London: Earthscan, 1994).

52. Lothar Brock, "Environmental Conflict Research—Paradigms and Perspectives," in Carius and Lietzmann, *Environmental Change and Security,* 45.

53. Lietzmann and Vest, *Environment and Security in an International Context.*

54. Frank Biermann, "Syndromes of Global Change: A Taxonomy for Peace and Conflict Research," in Carius and Lietzmann, *Environmental Change and Security,* 131–46.

55. Baechler, *Violence through Environmental Discrimination,* 236, 235.

56. Ibid., 235.

57. Ibid., 236.

58. Mohamed Suliman also ends his edited volume with a short conclusion meditating on the same "solution" in "Conflict Resolution among the Borana and the Fur: Similar Features, Different Outcomes," in Suliman, *Ecology, Politics, and Violent Conflict,* 286–90.

59. This also supports Elinor Ostrom's long-standing finding that local solutions that are pragmatic and refuse simplistic ideological formulas whether in support of markets or states are likely to deal with resource conflicts best. See Elinor Ostrom, *Governing the Commons: The Evolution of Institutions for Collective Action* (Cambridge: Cambridge University Press, 1990). More recent practical examples of watershed management are in Fiona Hinchcliffe, John Thompson, Jules Pretty, Irene Gujit, and Parmash Shah, eds., *Fertile Ground: The Impacts of Participatory Watershed Management* (London: Intermediate Technology Publications, 1999).

60. I have made this argument in regard to Canadian foreign policy in Simon Dalby, "Canadian National Security and Global Environmental Change," in J. Hanson and S. McNish, eds., *Canadian Strategic Forecast 1997—Canada and the World; Non-Traditional Security Threats* (Toronto: Canadian Institute for Strategic Studies, 1997), 19–40.

4. GEOPOLITICS AND HISTORY

1. In the preface to the reprinted book version of the essay in 2000, Kaplan notes that the assumption that what was happening in Africa would follow elsewhere was subsequently disproven. See Robert Kaplan, *The Coming Anarchy: Shattering the Dreams of the Post–Cold War* (New York: Random House, 2000), xiii.

2. Ronnie D. Lipschutz, "Environmental Conflict and Environmental Determinism: The Relative Importance of Social and Natural Factors," in N. P. Gleditsch, ed., *Conflict and the Environment* (Dordrecht: Kluwer, 1997),

35–50; Daniel Deudney, "Bringing Nature Back In: Geopolitical Theory from the Greeks to the Global Era," in Daniel Deudney and Richard Matthew, eds., *Contested Grounds: Security and Conflict in the New Environmental Politics* (Albany: State University of New York Press, 1999), 25–57.

3. Bruce Russett, *Grasping the Democratic Peace: Principles for a Post–Cold War World* (Princeton, N.J.: Princeton University Press, 1993).

4. Nils Petter Gleditsch, "Environmental Conflict and the Democratic Peace," in Gleditsch, ed., *Conflict and the Environment*, 91–106; M. I. Midlarsky, "Democracy and the Environment: An Empirical Assessment," *Journal of Peace Research* 35, no. 3 (1998): 341–61; and in general, Paul F. Diehl and Nils Petter Gleditsch, eds., *Environmental Conflict* (Boulder, Colo.: Westview, 2001).

5. I. Oren and J. Hays, "Democracies Rarely Fight One Another, but Developed Socialist States Rarely Fight at All," *Alternatives* 22, no. 4 (1997): 493–521.

6. John Agnew, "The Territorial Trap: The Geographical Assumptions of International Relations Theory," *Review of International Political Economy* 1, no. 1 (1994): 53–80; the argument also appears in John Agnew and Stuart Corbridge, *Mastering Space: Hegemony, Territory, and International Political Economy* (London: Routledge, 1995); and in John Agnew, *Geopolitics: Revisioning World Politics* (London: Routledge, 1998).

7. Robert O. Keohane, "Analyzing the Effectiveness of International Environmental Institutions," in Robert O. Keohane and Marc A. Levy, eds., *Institutions for Environmental Aid: Pitfalls and Promise* (Cambridge: MIT Press, 1996), 3.

8. Michael Redclift, *Wasted: Counting the Costs of Global Consumption* (London: Earthscan, 1996); Michael Redclift, "Environmental Security and Competition for the Environment," in Steve C. Lonergan, ed., *Environmental Change, Adaptation, and Security* (Dordrecht: Kluwer, 1999), 3–16.

9. John Vogler, *The Global Commons: Environmental and Technological Governance* (London: Wiley, 2000).

10. Partha Chatterjee and Matthias Finger, *The Earth Brokers: Power, Politics, and World Development* (London: Routledge, 1994); S. Chubin, "The South and the New World Order," in B. Roberts, ed., *Order and Disorder after the Cold War* (Cambridge: MIT Press, 1996), 429–49.

11. Mike Davis, *Ecology of Fear: Los Angeles and the Imagination of Disaster* (New York: Vintage, 1999), 16.

12. See B. L. Turner et al., *The Earth as Transformed by Human Action* (Cambridge: Cambridge University Press, 1990); also, from a rapidly growing literature, I. G. Simmons, *Environmental History: A Concise Introduction* (Oxford: Blackwell, 1993); and A. M. Mannion, *Global Environmental Change: A Natural and Cultural Environmental History* (London: Addison Wesley Longman, 1997). On climate specifically, see H. H. Lamb, *Climate,*

History, and the Modern World, 2d ed. (London: Routledge, 1995); and on the twentieth century, J. R. McNeill, *Something New under the Sun: An Environmental History of the Twentieth-Century World* (New York: Norton, 2000).

13. Richard Grove, *Ecology, Climate, and Empire: Colonialism and Global Environmental History, 1400–1940* (Cambridge, England: White Horse, 1997), 183.

14. See Evan Eisenberg, *The Ecology of Eden* (New York: Knopf, 1998).

15. Grove, *Ecology, Climate, and Empire,* 185.

16. On Kenya, colonization, the strategy of spatial divisions, and the establishment of reserve territories, see Roxanne Lynn Doty, *Imperial Encounters: The Politics of Representation in North-South Relations* (Minneapolis: University of Minnesota Press, 1996).

17. James Scott, *Seeing like a State: How Certain Schemes to Improve the Human Condition Have Failed* (New Haven, Conn.: Yale University Press, 1998).

18. Brian Fagan, *Floods, Famines, and Emperors: El Niño and the Fate of Civilizations* (New York: Basic Books, 1999). Australia is especially intriguing here; see Tim Flannery, *The Future Eaters: An Ecological History of the Australasian Lands and People* (New York: George Braziller, 1995).

19. H. R. Alker and P. M. Haas, "The Rise of Global Ecopolitics," in Nazli Choucri, ed., *Global Accord: Environmental Challenges and International Responses* (Cambridge: MIT Press, 1993), 133–71.

20. V. I. Vernadsky, "The Biosphere and the Noosphere," *American Scientist* 33 (1945): 1–12; F. Braudel, *Civilization and Capitalism, 15th–18th Century,* 3 vols. (New York: Harper & Row, 1981–1984).

21. F. Ratzel, *Politische Geographie* (Munich: Oldenbourg, 1897).

22. World Commission on Environment and Development, *Our Common Future* (Oxford: Oxford University Press, 1987).

23. James E. Lovelock, *The Ages of Gaia: A Biography of Our Living Earth* (New York: Norton, 1988).

24. Marc Levy, "Is the Environment a National Security Issue?" *International Security* 20 (1995): 35–62.

25. Intergovernmental Panel on Climate Change, *Climate Change: The IPCC Scientific Assessment* (Cambridge: Cambridge University Press, 1992); R. T. Watson, M. Zinyowera, and R. H. Moss, eds., *Climate Change 1995: Impacts, Adaptations, and Mitigation of Climate Change: Scientific-Technical Analyses* (Cambridge: Cambridge University Press, 1996).

26. R. B. Alley and M. L. Bender, "Greenland Ice Cores: Frozen in Time," *Scientific American* 278, no. 2 (February 1998): 80–85; Peter Bunyard, "How Climate Change Could Spin out of Control," *Ecologist* 29, no. 2 (1999): 68–75.

27. Lewis Mumford, *Technics and Civilization* (New York: Harcourt,

Brace, and World, 1934). Mumford's point was that industrial capitalism was fueled by coal from rocks of the Carboniferous period. Given current concerns with carbon dioxide and climate change, the term is even more evocative, even if petroleum doesn't come from Carboniferous era rocks!

28. An alternative formulation of these processes in terms of the rise of "Industria" is suggested in William Hipwell, "Industria, the Fourth World, and the Question of Territory," *Middle States Geographer* 30 (1997): 1–10.

29. Julian Saurin, "International Relations, Social Ecology, and the Globalization of Environmental Change," in Mark Imber and John Vogler, eds., *Environment and International Relations* (London: Routledge, 1996), 77–98; Peter Doran, "Upholding the 'Island of High Modernity': The Changing Climate of American Foreign Policy," in Paul G. Harris, ed., *Climate Change and American Foreign Policy* (New York: St. Martin's, 2000), 51–70; Matthew Paterson, "Car Culture and Global Environmental Politics," *Review of International Studies* 26, no. 2 (2000): 253–70.

30. Marvin S. Soroos, *The Endangered Atmosphere: Preserving a Global Commons* (Columbia: University of South Carolina Press, 1997).

31. Jared Diamond, *Guns, Germs, and Steel: The Fates of Human Societies* (New York: Norton, 1997).

32. Alfred Crosby, *Ecological Imperialism: The Biological Expansion of Europe 900–1900* (Cambridge: Cambridge University Press, 1986).

33. Eric Wolf, *Europe and the People without History* (Berkeley: University of California Press, 1982).

34. See the classic account in Suzanna Hecht and Alexander Cockburn, *The Fate of the Forest: Developers, Destroyers, and Defenders of the Amazon* (Harmondsworth, England: Penguin, 1990).

35. Michael Shapiro, *Violent Cartographies: Mapping Cultures of War* (Minneapolis: University of Minnesota Press, 1997).

36. R. Howitt, J. Connell, and P. Hirsch, eds., *Resources, Nations, and Indigenous Peoples* (Melbourne: Oxford University Press, 1996); B. R. Johnston, ed., *Who Pays the Price? The Sociocultural Context of Environmental Crisis* (Washington, D.C.: Island Press, 1994).

37. Neil Smith, *Uneven Development: Nature, Capital, and the Production of Space* (Oxford: Blackwell, 1990).

38. Obviously the process is more complicated than this; indigenous populations do have effects on the land, and colonization has historically often both enmeshed them in trading relationships and forced them to move. Simple before and after comparisons miss these important dimensions of change. See William Cronon, *Changes in the Land: Indians, Colonists, and the Ecology of New England* (New York: Hill and Wang, 1983); and more generally, Lisa M. Benton and John Rennie Short, *Environmental Discourse and Practice* (Oxford: Blackwell, 1999), 27–59.

39. Akhil Gupta and James Ferguson, "Beyond 'Culture': Space, Identity,

and the Politics of Difference," in Akhil Gupta and James Ferguson, eds., *Culture, Power, Place: Explorations in Critical Anthropology* (Durham, N.C.: Duke University Press, 1997), 33–51.

40. M. Gadgil and R. Guha, *Ecology and Equity: The Use and Abuse of Nature in Contemporary India* (London: Routledge, 1995).

41. A. Sachs, *Eco-Justice: Linking Human Rights and the Environment*, Worldwatch Paper 127 (Washington, D.C.: Worldwatch Institute, 1995); Michael Watts, "Nature as Artifice and Artifact," in Bruce Braun and Noel Castree, eds., *Remaking Reality: Nature at the Millennium* (London: Routledge, 1998), 243–68.

42. Daniel Botkin, *Discordant Harmonies: A New Ecology for the Twenty-first Century* (Oxford: Oxford University Press, 1990).

43. Matthias Finger, "The Military, the Nation State, and the Environment," *Ecologist* 21, no. 5 (1991): 220–25.

44. Marcus Doel, *Poststructuralist Geographies* (Edinburgh: University of Edinburgh Press, 1999).

45. Ken Conca, "Rethinking the Ecology-Sovereignty Debate," *Millennium* 23, no. 3 (1994): 701–11.

46. On the complexity of this in China in the period of reform and marketization in the 1990s, see J. S. S. Muldavin, "Environmental Degradation in Heilongjiang: Policy Reform and Agrarian Dynamics in China's New Hybrid Economy," *Annals of the Association of American Geographers* 87, no. 4 (1997): 579–613.

47. On recent non–green revolution innovations in food production beyond the control of agricultural corporations, see Gordon Prain, Sam Fujisaka, and Michael D. Warren, eds., *Biological and Cultural Diversity: The Role of Indigenous Agricultural Experimentation in Development* (London: Intermediate Technology Publications, 1999).

48. Given the cavalier disregard for these matters in Robert Kaplan's account of West Africa, Melissa Leach and James Fairhead's analysis in their "Challenging Neo-Malthusian Deforestation Analyses in West Africa's Dynamic Forest Landscapes" (*Population and Development Review* 26, no. 1 [2000]: 17–43) is especially instructive.

49. J. M. Silliman and Y. King, eds., *Dangerous Intersections: Feminist Perspectives on Population, Environment, and Development* (Boston: South End 1999).

5. IMPERIAL LEGACIES, INDIGENOUS LIVES

1. Karl Polanyi, *The Great Transformation: The Political and Economic Origins of our Time* (Boston: Beacon, 1957).

2. Richard Grove, *Ecology, Climate, and Empire: Colonialism and Global Environmental History, 1400–1940* (Cambridge, England: White Horse, 1997).

3. See O. Wæver, B. Buzan, M. Kelstrup, and P. Lemaitre, *Identity, Migration, and the New Security Agenda in Europe* (London: Pinter, 1993).

4. M. Connelly and P. Kennedy, "Must It Be the West against the Rest?" *Atlantic Monthly* 274, no. 6 (1994): 61–83.

5. Ibid., 62.

6. This image supports the overall argument in the article in many ways. It is a domestic enclosure that might be read in the gendered terms of the feminization of modern consumption or in terms of cooking as the mark of civilization. The white picket fence has long been a cultural marker of middle-class status aspiration, contrasted here to the proletarian "mob." Many of the faces outside the fence are obviously younger than the home owner inside, suggesting the demographic contrast between white races and many populations in the global South. Thanks to both Mathew Coleman and Matthew Paterson for their comments here.

7. Connelly and Kennedy, "Must It Be," 69–70.

8. On "optimistic" rejoinders to the Kaplan thesis, see A. E. Server, "The End of the World Is Nigh—Or Is It?" *Fortune*, 2 May 1994, 123–24; M. Gee, "Surprise! The World Gets Better!" *World Press Review* 14, no. 7 (July 1994): 18–20.

9. See Richard J. Barnet and J. Cavanagh, *Global Dreams: Imperial Corporations and the New World Order* (New York: Simon and Schuster, 1994).

10. Paul Kennedy, *Preparing for the Twenty-First Century* (New York: Harper Collins, 1993).

11. Virginia Abernethy, "Optimism and Overpopulation," *Atlantic Monthly* 274, no. 6 (1994): 84–91; and in general, V. Abernethy, *Population Politics: The Choices That Shape Our Future* (New York: Plenum Press, 1993). Although her argument can be criticized because it presumes social stability and patterns of reproductive behaviors that may not be appropriate assumptions for the poorest areas of the world, it does offer an alternative justification for modest grassroots aid projects aimed at the poorest sections of society. On how impoverishment of environmental resources can lead to increases rather than decreases in family size, see P. S. Dasgupta, "Population, Poverty, and the Local Environment," *Scientific American* 272, no. 2 (1995): 40–45.

12. Michael Shapiro, "Narrating the Nation, Unwelcoming the Stranger: Anti-Immigration Policy in Contemporary 'America,'" *Alternatives* 22, no. 1 (1997): 1–34.

13. David Campbell, *Politics without Principle: Sovereignty, Ethics, and the Narratives of the Gulf War* (Boulder, Colo.: Lynne Rienner, 1993).

14. See Anthony Richmond, *Global Apartheid: Refugees, Racism, and the New World Order* (Toronto: Oxford University Press, 1994). Thomas Schelling argues that South Africa is an appropriate model for the global polity, geographically divided as it is between rich and poor, in "The Global Dimension," in G. Allison and G. Treverton, eds., *Rethinking America's Security: Beyond Cold War to New World Order* (New York: Norton, 1992), 196–210. More generally, see "Against Global Apartheid," special issue of *Alternatives* 19, no. 2 (1994); G. Kohler, "The Three Meanings of Global Apartheid: Empirical, Normative, Existential," *Alternatives* 20, no. 3 (1995): 403–13; Simon Dalby, "Globalization or Global Apartheid? Boundaries and Knowledge in Postmodern Times," *Geopolitics* 3, no. 1 (1998): 132–50.

15. Robert Ford, "The Population-Environment Nexus and Vulnerability Assessment in Africa," *GeoJournal* 35 (1995): 207–16.

16. Piers Blaikie and Harold Brookfield, *Land Degradation and Society* (London: Methuen, 1987); Richard Peet and Michael Watts, eds., *Liberation Ecologies: Environment, Development, Social Movements* (New York: Routledge, 1996); Raymond Bryant and Sinead Bailey, *Third World Political Ecology* (London: Routledge, 1997); Roger Keil, David V. J. Bell, Peter Penz, and Leesa Fawcett, eds., *Political Ecology: Global and Local* (London and New York: Routledge, 1998); Philip Stott and Sian Sullivan, eds., *Political Ecology: Science, Myth, and Power* (London: Arnold, 2000); Nancy L. Peluso and Michael Watts, eds., *Violent Environments* (Ithaca, N.Y.: Cornell University Press, 2001).

17. D. Rocheleau, B. Thomas-Slayter, and E. Wangari, eds., *Feminist Political Ecology: Global Issues and Local Experiences* (New York: Routledge, 1996).

18. N. L. Peluso, "Coercing Conservation: The Politics of State Resource Control," in R. D. Lipschutz and K. Conca, eds., *The State and Social Power in Global Environmental Politics* (New York: Columbia University Press, 1993), 46–70.

19. J. Swift, ed., "War and Rural Development in Africa," special issue of *Institute of Development Studies Bulletin* 27, no. 3 (1996). More generally, see William Reno, *Warlord Politics and African States* (Boulder, Colo.: Lynne Rienner, 1998).

20. See Nicholas Xenos, *Scarcity and Modernity* (London: Routledge, 1989).

21. Indra de Soysa and Nils Petter Gleditsch, *To Cultivate Peace—Agriculture in a World of Conflict,* Report 1/99 (Oslo: International Peace Research Institute, 1999), 35.

22. These connections are discussed in terms of "economic genocide" in Michel Chossudovsky, "Human Security and Economic Genocide in Rwanda," in Caroline Thomas and Peter Wilkin, eds., *Globalization, Human*

Security, and the African Experience (Boulder, Colo.: Lynne Rienner, 1999), 117–26; for a comprehensive treatment of the inadequacies of Malthusian frameworks in this case, see Leif Ohlsson, *Environment, Scarcity, and Conflict: A Study of Malthusian Concerns* (Goteburg: Goteburg University Department of Peace and Development Research, 1999).

23. P. Blaikie, T. Cannon, I. Davis, and B. Wisner, *At Risk: Natural Hazards, People's Vulnerability, and Disasters* (London: Routledge, 1994). The case of affluent Californians who choose to live in dangerous areas is obviously related more to their lack of environmental knowledge, inadequate land use planning, and their apparent assumption that nature ought to be bent to human will regardless of the cost. See Mike Davis, *Ecology of Fear: Los Angeles and the Imagination of Disaster* (New York: Vintage, 1999).

24. Anthony Oliver-Smith and Susanna M. Hoffman, *The Angry Earth: Disaster in Anthropological Perspective* (London: Routledge, 1999).

25. L. Pettiford, "Towards a Redefinition of Security in Central America: The Case of Natural Disasters," *Disasters* 19 (1995): 148–55.

26. R. K. Molvoer, "Environmentally Induced Conflicts?" *Bulletin of Peace Proposals* 22 (1991): 175–88.

27. Indra de Soysa, "Natural Resources and Civil War: Shrinking Pie or Honey Pot?" Paper presented to the annual convention of the International Studies Association, Los Angeles, March 2000.

28. Indra de Soysa, "The Resource Curse: Are Civil Wars Driven by Rapacity or Paucity?" in Mats Berdal and David M. Malone, eds., *Greed and Grievance: Economic Agendas in Civil Wars* (Boulder, Colo.: Lynne Rienner, 2000), 113–35. See also David Keen, *The Economic Functions of Violence in Civil Wars*, Adelphi Paper 320 (London: International Institute for Strategic Studies, 1998).

29. These relationships are tabulated in Edward B. Barbier and Thomas Homer-Dixon, "Environmental Change, Social Conflict, and Limits for Adaptation in Developing Countries," in Steve C. Lonergan, ed., *Environmental Change, Adaptation, and Security* (Dordrecht: Kluwer, 1999), 335–47.

30. Eric B. Ross, *The Malthus Factor: Poverty, Politics, and Population in Capitalist Development* (London: Zed, 1998).

31. Adam Hochschild, *King Leopold's Ghost: A Story of Greed, Terror, and Heroism in Colonial Africa* (Boston: Houghton Mifflin, 1998).

32. Kirk Talbott and Melissa Brown, "Forest Plunder in South East Asia: An Environmental Security Nexus in Burma and Cambodia," *Environmental Change and Security Project* 4 (1998): 53.

33. Ibid., 59.

34. Colin H. Kahl, "Population Growth, Environmental Degradation, and State Sponsored Violence: The Case of Kenya, 1991–93," *International Security* 23, no. 3 (1998): 80–119; Kahl, "The Political Ecology of Conflict:

Lessons from the Philippines," paper presented to the annual convention of the International Studies Association, Los Angeles, March 2000.

35. Philippe Le Billon, "The Political Ecology of War: Natural Resources and Armed Conflicts," *Political Geography* 20 (2001): 561–84.

36. S. Jewitt, "Europe's 'Others'? Forestry Policy and Practices in Colonial and Postcolonial India," *Environment and Planning D: Society and Space* 13 (1995): 67–90.

37. Thom Kuehls, *Beyond Sovereign Territory: The Space of Ecopolitics* (Minneapolis: University of Minnesota Press, 1996).

38. A. Pinheiro and P. Cesar, "A Vision of the Brazilian National Security Policy on the Amazon," *Low Intensity Conflict and Law Enforcement* 3 (1994): 387–409.

39. A. Gedicks, *The New Resource Wars: Nature and Environmental Struggles against Multinational Companies* (Boston: South End, 1993); B. R. Johnston, ed., *Who Pays the Price? The Sociocultural Context of Environmental Crisis* (Washington, D.C.: Island Press, 1994).

40. The classic statement on this theme has become *Ecologist: Whose Common Future? Reclaiming the Commons* (Philadelphia: New Society, 1993). For a recent update linking the argument to intellectual property rights, see Marian A. L. Miller, "Tragedy of the Commons: The Enclosure and Commodification of Knowledge," in Dimitris Stevis and Valerie Assetto, eds., *The International Political Economy of the Environment* (Boulder, Colo.: Lynne Rienner, 2001), 111–34.

41. N. L. Whitehead and R. B. Ferguson, eds., *War in the Tribal Zone: Expanding States and Indigenous Warfare* (Santa Fe, N.M.: School of American Research, distributed by University of Washington Press, 1992).

42. H. Rangan, "From Chipko to Uttaranchal: Development, Environment, and Social Protest in the Garhwal Himalayas, India," in Peet and Watts, *Liberation Ecologies*, 205–26.

43. Bernard Nietschmann, "The Fourth World: Nations versus States," in G. J. Demko and W. B. Wood, eds., *Reordering the World: Geopolitical Perspectives on the Twenty-first Century* (Boulder, Colo.: Westview, 1994).

44. Alfred Crosby, *Ecological Imperialism: The Biological Expansion of Europe 900–1900* (Cambridge: Cambridge University Press, 1986).

45. J. Masco, "States of Insecurity: Plutonium and Post–Cold War Anxiety in New Mexico, 1992–96," in J. Weldes, M. Laffey, H. Gusterson, and R. Duvall, eds., *Cultures of Insecurity: States, Communities, and the Production of Danger* (Minneapolis: University of Minnesota Press, 1999), 203–31.

46. D. L. Fixico, *The Invasion of Indian Country in the Twentieth Century: American Capitalism and Tribal National Resources* (Niwat: University Press of Colorado, 1998).

47. For an ethnography of the Mi'kmaq, see Stephen A. Davis, *Mi'kmaq* (Halifax, Nova Scotia: Nimbus, 1997). The more recent stories of dispos-

session and assimilation are told in Rita Joe and Lesley Choyce, eds., *The Mi'kmaq Anthology* (East Lawrencetown, Nova Scotia: Pottersfield Press, 1997).

48. Alf Hornborg, "The Mi'kmaq of Nova Scotia: Environmentalism, Ethnicity, and Sacred Places," in Alf Hornborg and Mikael Kurkiala, eds., *Voices of the Land: Identity and Ecology in the Margins* (Lund, Sweden: Lund University Press, 1998), 135–72; Alf Hornborg, "Mi'kmaq Environmentalism: Local Incentives and Global Projections," in L. A. Sandberg and S. Sörlin, eds., *Sustainability the Challenge: People, Power, and the Environment* (Montreal: Black Rose Books, 1998), 202–11.

49. Winona LaDuke, *All Our Relations: Native Struggles for Land and Life* (Cambridge, Mass.: South End, 1999).

50. Field interview 1997. This interview was part of a research project conducted on "Community, Identity, and Environmental Threat." See Simon Dalby and Fiona Mackenzie, "Reconceptualizing Local Community: Environment, Identity, and Threat," *Area* 29, no. 2 (1997): 99–108.

51. Ward Churchill, *Struggle for the Land: Native North American Resistance to Genocide, Ecocide, and Colonization* (Winnipeg: Arbeiter Ring, 1999), 367.

52. Field interview, 1997.

53. Richard Grove, *Green Imperialism: Colonial Expansion, Tropical Island Edens, and the Origins of Environmentalism, 1600–1800* (Cambridge: Cambridge University Press, 1995).

54. On this point, see Tim W. Luke, *Capitalism, Democracy, and Ecology: Departing from Marx* (Champaign: University of Illinois Press, 1999), 118–42.

55. Arturo Escobar, "Constructing Nature: Elements of a Post-Structural Political Ecology," in Peet and Watts, *Liberation Ecologies, 50.*

6. SHADOWS, FOOTPRINTS, AND ENVIRONMENTAL SPACE

1. Raymond Bryant and Sinead Bailey, *Third World Political Ecology* (London: Routledge, 1997); Michael Redclift, *Wasted: Counting the Costs of Global Consumption* (London: Earthscan, 1996).

2. R. D. Lipschutz, *When Nations Clash: Raw Materials, Ideology, and Foreign Policy* (Cambridge, Mass.: Ballinger, 1989). See also S. D. Krasner, *Defending the National Interest: Raw Materials Investments and U.S. Foreign Policy* (Princeton, N.J.: Princeton University Press, 1978).

3. See Miquel de Larrinaga, "(Re)politicizing the Discourse: Globalization Is a S(h)ell Game," *Alternatives* 25, no. 2 (2000): 145–82; and Volker Boge, "Mining, Environmental Degradation, and War: The Bougainville

Case," in M. Suliman, ed., *Ecology, Politics, and Violent Conflict* (London: Zed, 1999), 211–27.

4. John Vogler, *The Global Commons: Environmental and Technological Governance* (London: Wiley, 2000).

5. The inadequacies of geographical reasoning are key themes in Martin W. Lewis and Karen E. Wigen, *The Myth of Continents: A Critique of Metageography* (Berkeley: University of California Press, 1997); and John Agnew, *Geopolitics: Revisioning World Politics* (London: Routledge, 1998).

6. Nils Petter Gleditsch, "Armed Conflict and the Environment: A Critique of the Literature," *Journal of Peace Research* 35, no. 3 (1998): 381–400.

7. J. MacNeill, P. Winsemius, and T. Yakushiji, *Beyond Interdependence* (New York: Oxford University Press, 1991), 58–59.

8. Peter Dauvergne, *Shadows in the Forest: Japan and the Politics of Timber in South East Asia* (Cambridge: MIT Press, 1997), 11.

9. Ibid., 166.

10. N. Marinova, "Indonesia's Fiery Crises," *Journal of Environment and Development* 8, no. 1 (1999): 70–81.

11. A loosely similar, earlier formulation of these themes in terms of "phantom carrying capacity" and calculations of "fossil fuel acreage" is in William R. Catton Jr., *Overshoot: The Ecological Basis of Revolutionary Change* (Urbana: University of Illinois Press, 1980).

12. M. Wackernagel and W. Rees, *Our Ecological Footprint: Reducing Human Impact on the Earth* (Philadelphia and Gabriola Island, British Columbia: New Society, 1996), 9.

13. Ibid., 11.

14. Ibid. Carrying capacity in these terms is discussed in detail in W. E. Rees, "Revisiting Carrying Capacity: Area Based Indicators of Sustainability," *Population and Environment* 17, no. 3 (1996): 195–215.

15. Wackernagel and Rees, *Our Ecological Footprint*, 85.

16. W. E. Rees and M. Wackernagel, "Ecological Footprints and Appropriated Carrying Capacity: Measuring the Natural Capital Requirements of the Human Economy," in A.-M. Jannson, M. Hammer, C. Folke, and R. Constanza, eds., *Investing in Natural Capital: The Ecological Economics Approach to Sustainability* (Washington, D.C.: Island Press, 1994), 378.

17. Wackernagel and Rees, *Our Ecological Footprint*, 88.

18. Ibid., 89.

19. W. E. Rees, "Ecological Footprints and Appropriated Carrying Capacity: What Urban Economics Leaves Out," *Environment and Urbanization* 4, no. 2 (1992): 128.

20. W. Sachs, R. Loske, and Manfred Linz, *Greening the North: A Post-Industrial Blueprint for Ecology and Equity* (London: Zed, 1998).

21. Ibid., 12. They cite J. B. Opschoor, *Environment, Economy, and Sus-*

tainable Development (Groningen, The Netherlands: Wolters-Noordhoff, 1992); and J. B. Opschoor, *Economic Incentives and Environmental Policies* (Dordrecht: Kluwer Academic, 1994) as the original sources of the concept.

22. D. H. Meadows, D. L. Meadows, J. Randers, and W. W. Behrens III, *The Limits to Growth* (London: Pan, 1974).

23. Sachs, Loske, and Linz, *Greening the North*, 12–13.

24. Ibid., 14. This is a rearticulation of classic liberal rights to use property as long as doing so does not harm others' similar rights. This principle was the basis for international environmental agreements in the 1972 United Nations Stockholm Conference on the Human Environment. This reasserted state sovereignty, allowing for the development and use of natural resources in a manner that did not cause damage to other states.

25. On the difficulties of universal claims to justice on a variegated planetary surface, see I. Wallace and D. Knight, "Societies in Space and Place," in F. O. Hampson and J. Reppy, eds., *Earthly Goods: Environmental Change and Social Justice* (Ithaca, N.Y.: Cornell University Press, 1996), 75–95.

26. Sachs, Loske, and Linz, *Greening the North*, 70.

27. A. Sachs, *Eco-Justice: Linking Human Rights and the Environment*, Worldwatch Paper 127 (Washington, D.C.: Worldwatch Institute, 1995).

28. Henry Shue, "Global Environment and International Inequality," *International Affairs* 75, no. 3 (1999): 531–45; Ferenc L. Toth, ed., *Fair Weather: Equity Concerns in Climate Change* (London: Earthscan, 1999).

29. See the overview of atmospheric agreements in Marvin S. Soroos, *The Endangered Atmosphere: Preserving a Global Commons* (Columbia: University of South Carolina Press, 1997).

30. Arguments about the requirements of all states to do something by way of greenhouse gas reductions were used repeatedly in the United States to delay American initiatives in the 1990s. See Paul G. Harris, ed., *Climate Change and American Foreign Policy* (New York: St. Martin's, 2000).

31. Anil Agarwal and Sunita Narain, *Global Warming in an Unequal World: A Case of Environmental Colonialism* (New Delhi: Centre for Science and Environment, 1991). The essential point of this argument is captured in the Scott Willis cartoon on the front cover of this report. It shows a diminutive peasant about to chop down a tree. However, just prior to the first blow of the axe, a large male figure standing in an enormous automobile, with license plate reading "developed countries" and with actively emitting exhaust pipe prominently displayed, remonstrates with the peasant, saying, "Yo! Amigo!! you can't cut that tree. We need it to stop the greenhouse effect."

32. I have incorporated these points as part of a larger "Southern critique" of the environmental security discourse in my "Threats from the South?" in Daniel Deudney and Richard Matthew, eds., *Contested Grounds: Security*

and Conflict in the New Environmental Politics (Albany: State University of New York Press, 1999), 155–85.

33. Thomas E. Downing, ed., *Climate Change and World Food Security* (Berlin: Springer-Verlag, 1996).

34. Gunther Baechler, *Violence through Environmental Discrimination: Causes, Rwanda Arena, and Conflict Model* (Dordrecht: Kluwer, 1999).

35. See, in general, M. Watts and D. Goodman, *Globalizing Food: Agrarian Questions and Global Restructuring* (London: Routledge, 1997); and as an example of commodity chains, Robert N. Gwynne, "Globalization, Commodity Chains, and Fruit Exporting Regions in Chile," *TESG: Tijdschrift voor Economische en Sociale Geografie* 90, no. 2 (1999): 211–25.

36. Sachs, Loske, and Linz, *Greening the North,* 80.

37. Valerie Percival and T. Homer-Dixon, "Environmental Scarcity and Violent Conflict: The Case of South Africa," *Journal of Peace Research* 35, no. 3 (1998): 279–98.

38. For an overview of the environmental context, see Southern African Research and Documentation Centre, *State of the Environment in Southern Africa* (Harare: Southern African Research and Documentation Centre, 1994).

39. Larry A. Swatuk and Peter Vale, "Why Democracy Is Not Enough: Southern Africa and Human Security in the Twenty-first Century," *Alternatives* 24, no. 3 (1999): 364.

40. Ibid.

41. J. Warburton-Lee, "Breaking Down the Barricades," *Geographical Magazine* 71, no. 9 (1999): 18–25.

42. See Fikret Berkes and Carl Folke, eds., *Linking Social and Ecological Systems: Management Practices and Social Mechanisms for Building Resilience* (Cambridge: Cambridge University Press, 1998).

43. For this critique of Thomas Homer-Dixon's framework, see David A. McDonald, "Lest the Rhetoric Begin: Migration, Population, and the Environment in Southern Africa," *Geoforum* 30 (1999): 13–25.

44. Sachs, Loske, and Linz, *Greening the North,* 158.

45. Ibid., 159.

46. On circular reasoning, see Gwyn Prins, "Politics and the Environment," in Prins, ed., *Threats without Enemies: Facing Environmental Insecurity* (London: Earthscan, 1993), 171–91.

47. Sachs, Loske, and Linz, *Greening the North,* 136.

48. Not all agriculture in the South necessarily dispossesses local subsistence farmers, appropriates scarce water supplies, or directly contributes to soil erosion. However, the fossil fuel consumption in these practices is considerable. See Hazel R. Barrett, Brian W. Illbery, Angela W. Browne, and Tony Binns, "Globalization and the Changing Networks of Food Supply: The Im-

portation of Fresh Horticultural Produce from Kenya into the UK," *Transactions of the Institute of British Geographers*, n.s., 24 (1999): 159–74.

49. Richard A. Shroeder and Krisnawati Suryanata, "Gender and Class Power in Agroforestry Systems," in Richard Peet and Michael Watts, eds., *Liberation Ecologies: Environment, Development, Social Movements* (New York: Routledge, 1996), 188–204.

50. Jerry Mander and Edward Goldsmith, eds., *The Case against the Global Economy: And for a Turn to the Local* (San Francisco: Sierra Club Books, 1996).

51. Braden R. Allenby, "Environmental Security: Concept and Implementation," *International Political Science Review* 21, no. 1 (2000): 14.

7. ECOLOGICAL METAPHORS OF SECURITY

1. On containers, physics metaphors, and security, see Paul Chilton, *Security Metaphors: Cold War Discourse from Containment to Common House* (New York: Peter Lang, 1996).

2. J. Ann Tickner, *Gender in International Relations: Feminist Perspectives on Achieving Global Security* (New York: Columbia University Press, 1992); Tickner, "Revisioning Security," in Ken Booth and Steve Smith, eds., *International Relations Theory Today* (Cambridge: Polity, 1995), 175–97.

3. Eric Laferrière, "Emancipating International Relations Theory: An Ecological Perspective," *Millennium* 25, no. 1 (1996): 53–75. Neither have these themes been followed up in the discussions of ecology and sovereignty; see Karen Litfin, "Sovereignty in World Ecopolitics," *Mershon International Studies Review* 41 (1997): 167–204. While works such as Hayward Alker and Peter Haas, "The Rise of Global Ecopolitics," in Nazli Choucri, ed., *Global Accord: Environmental Challenges and International Responses* (Cambridge: MIT Press, 1993); Julian Saurin, "Global Environmental Degradation, Modernity, and Environmental Knowledge," in Caroline Thomas, ed., *Rio: Unravelling the Consequences* (Ilford, England: Frank Cass, 1994); and Matthew Paterson, *Understanding Global Environmental Politics* (London: Macmillan, 2000), suggest taking environment seriously to investigate international relations, the theme has not been followed up in detail in regard to security.

4. For a discussion of security in these terms, see Richard James Blackburn, *The Vampire of Reason* (London: Verso, 1990).

5. For a recent rereading of Kenneth Waltz's classic treatment of Rousseau in environmental terms, see Thom Kuehls, "Between Sovereignty and Environment: An Exploration of the Discourse of Government," in Karen Litfin, ed., *The Greening of Sovereignty in World Politics* (Cambridge: MIT

Press, 1998), 31–53; and Kenneth Waltz, *Man, the State, and War* (New York: Columbia University Press, 1959).

6. Anna Bramwell, *Ecology in the Twentieth Century: A History* (New Haven, Conn.: Yale University Press, 1989).

7. It seems more likely that the individuals in the picture are protobourgeois men in need of a social contract to protect their property from each other's claims. In general, see Michael C. Williams, "Hobbes and International Relations: A Reconsideration," *International Organization* 50, no. 2 (1996): 213–36.

8. Kuehls, "Between Sovereignty and Environment"; David Bedford and Thom Workman, "The Great Law of Peace: Alternative Inter-Nation(al) Practices and the Iroquoian Confederacy," *Alternatives* 22, no. 1 (1997): 87–111.

9. See, for example, Clarence Glacken, *Traces on the Rhodian Shore* (Berkeley: University of California Press, 1967); and William Leiss, *The Domination of Nature* (Boston: Beacon, 1974).

10. This is related at the largest scale to a specifically Eurocentric understanding of the world; see J. M. Blaut, *The Colonizer's Model of the World* (New York: Guilford, 1993).

11. Harold Sprout and Margaret Sprout, *The Ecological Perspective on Human Affairs with Special Reference to International Politics* (Princeton, N.J.: Princeton University Press, 1965), 27. See also F. Kratochwil, "Of Systems, Boundaries, and Territoriality," *World Politics* 39, no. 1 (1986): 27–52.

12. The Sprouts also pointedly note that when the term is used loosely as in such formulations as "the international system's 'environment,'" it offers more confusion than intelligibility (Sprout and Sprout, *Ecological Perspective,* 39).

13. Ronnie Lipschutz, "The Nature of Sovereignty and the Sovereignty of Nature: Problematizing the Boundaries between Self, Society, State, and System," in Litfin, *The Greening of Sovereignty,* 109–38.

14. Chris Gibson, "Cartographies of the Colonial/Capitalist State: A Geopolitics of Indigenous Self-Determination in Australia," *Antipode* 31, no. 1 (1999): 45–79.

15. Jennifer Turpin and Lois Ann Lorentzen, eds., *The Gendered New World Order: Militarism, Development, and the Environment* (New York: Routledge, 1996).

16. Carolyn Merchant, *The Death of Nature: Women, Ecology, and the Scientific Revolution* (San Francisco: Harper and Row, 1980); Val Plumwood, *Feminism and the Mastery of Nature* (London: Routledge, 1993).

17. S. Bernstein, R. Ned Lebow, J. Gross Stein, and S. Webber, "God Gave

Physics the Easy Problems: Adapting Social Science to an Unpredictable World," *European Journal of International Relations* 6, no. 1 (2000): 43–76.

18. M. DeLanda, *War in the Age of Intelligent Machines* (New York: Zone Books, 1991).

19. Arthur Westing, ed., "Armed Conflict and Environmental Security," special edition of *Environment and Security* 1, no. 2 (1997).

20. Joni Seager, *Earth Follies: Coming to Feminist Terms with the Global Environmental Crisis* (New York: Routledge, 1993).

21. The pertinent literature is huge, but for a nonmathematical overview, see Robert E. Ulanowicz, *Ecology, the Ascendent Perspective* (New York: Columbia University Press, 1997).

22. Barry Commoner, *The Closing Circle* (New York: Knopf, 1971).

23. See, for instance, T. Miller, *Living in the Environment* (Belmont, Calif.: Wadsworth, 1990); and M. D. Morgan, J. M. Moran, and J. H. Wiersma, *Environmental Science* (Dubuque, Iowa: Wm. C. Brown, 1993).

24. Daniel Botkin, *Discordant Harmonies: A New Ecology for the Twenty-first Century* (Oxford: Oxford University Press, 1990).

25. G. A. DeLeo and S. Levin, "The Multifaceted Aspects of Ecosystem Integrity," *Conservation Ecology* [online] 1, no. 1 (1997): 3, at http://www.consecol.org/vol1/iss1/art3; James J. Kay, Henry A. Regier, Michele Boyle, and George Francis, "An Ecosystem Approach to Sustainability: Addressing the Challenge of Complexity," *Futures* 31 (1999): 721–42.

26. Alfred Crosby, *Ecological Imperialism: The Biological Expansion of Europe 900–1900* (Cambridge: Cambridge University Press, 1986).

27. Amita Baviskar, *In the Belly of the River: Tribal Conflicts over Development in the Narmada Valley* (Delhi: Oxford University Press, 1995).

28. Richard Falk, *On Humane Governance: Toward a New Global Politics* (University Park: Pennsylvania University Press, 1995).

29. James E. Lovelock, *Gaia: A New Look at Life on Earth* (Oxford: Oxford University Press, 1979); James E. Lovelock, *The Ages of Gaia: A Biography of Our Living Earth* (New York: Norton, 1988).

30. As Lovelock's eloquent "daisyworld" computer simulations have shown, this self-regulation can be easily demonstrated in mathematical models using simple feedback loops involving albedo-diverse floral communities on the surface of a planet.

31. The most comprehensive overview remains B. L. Turner et al., *The Earth as Transformed by Human Action* (Cambridge: Cambridge University Press, 1990).

32. R. Goodland, H. Daly, and H. Serafy, *Population, Technology, Lifestyle* (New York: Island Press, 1992); Thomas Prugh et al., *Natural Capital and Human Economic Survival* (Boca Raton, Fla.: Lewis, 1999).

33. R. F. Dasmann, "Towards a Biosphere Consciousness," in D. Worster, ed., *The Ends of the Earth: Perspective on Modern Environmental History* (Cambridge: Cambridge University Press, 1988), 177–88; Madhav Gadgil, "Prudence and Profligacy: A Human Ecological Perspective," in Timothy M. Swanson, ed., *The Economics and Ecology of Biodiversity Decline: The Forces Driving Global Change* (Cambridge: Cambridge University Press, 1995).

34. M. Goodman, ed., *Privatizing Nature: Political Struggles for the Global Commons* (New Brunswick, N.J.: Rutgers University Press, 1998).

35. M. Gadgil and R. Guha, *Ecology and Equity: The Use and Abuse of Nature in Contemporary India* (London: Routledge, 1995).

36. Michael Redclift, *Wasted: Counting the Costs of Global Consumption* (London: Earthscan, 1996).

37. The term "environmental refugee" has become commonplace in environmental discussions. See Patricia Saunders, "Environmental Refugees: The Origin of a Construct," in Philip Stott and Sian Sullivan, eds., *Political Ecology: Science, Myth and Power* (London: Arnold, 2000), 218–46. The Gadgil and Guha formulation emphasizes the displacement of populations by the expanding commercial sector, causing global degradation and population displacement (rather than the Malthusian formulation of peasants exhausting their local environments), which Saunders once again traces back to W. Vogt, *The Road to Survival* (New York: Sloan, 1948). (See chapter 2, note 18.)

38. Gadgil, "Prudence and Profligacy," 107.

39. P. J. Taylor, "World Cities and Territorial States under Conditions of Contemporary Globalization," *Political Geography* 19, no. 1 (2000): 5–32.

40. See Warren Magnusson, "Social Movements and the Global City," *Millennium* 23, no. 3 (1994): 621–45, reprinted as chapter 12 in Warren Magnusson, *The Search for Political Space* (Toronto: University of Toronto Press, 1996).

41. The Commission on Global Governance, *Our Global Neighbourhood* (Oxford: Oxford University Press, 1995).

42. Daniel Deudney, "Global Village Sovereignty: Intergenerational Sovereign Publics, Federal-Republican Earth Constitutions, and Planetary Identities," in Litfin, *The Greening of Sovereignty,* 299–325.

43. It is not much of a stretch to reread discussions of global civil society in these terms; see Paul Wapner, *Environmental Activism and World Civic Politics* (Albany: State University of New York Press, 1996); and Ronnie Lipschutz with Judith Mayer, *Global Civil Society and Global Environmental Governance: The Politics of Nature from Place to Planet* (Albany: State University of New York Press, 1996).

44. L. Allison, "On Dirty Public Things," *Political Geography Quarterly* 5 (1986): 241–51.

45. Daniel Faber, *Environment under Fire: Imperialism and Ecological Crisis in Central America* (New York: Monthly Review Press, 1993); Andrew Szasz, *Ecopopulism: Toxic Waste and the Movement of Environmental Justice* (Minneapolis: University of Minneapolis Press, 1994).

46. R. D. Bullard, *Dumping in Dixie: Race, Class, and Environmental Quality* (Boulder, Colo.: Westview, 1990).

47. B. R. Johnston, ed., *Who Pays the Price? The Sociocultural Context of Environmental Crisis* (Washington, D.C.: Island Press, 1994); Tom Athanasiou, *Divided Planet: The Ecology of Rich and Poor* (Boston: Little Brown, 1996).

48. Marian Miller, *The Third World in Global Environmental Politics* (Boulder, Colo.: Lynne Rienner, 1995).

49. M. Connelly and P. Kennedy, "Must It Be the West against the Rest?" *Atlantic Monthly* 274, no. 6 (1994): 61–83.

50. The arguments are summarized in Krishna B. Ghimire, "Parks and People: Livelihood Issues in National Parks Management in Thailand and Madagascar," in Dharam Ghai, ed., *Development and Environment: Sustaining People and Nature* (Oxford: Blackwell, 1994), 195–229.

51. See Jonathon S. Adams and Thomas O. McShane, *The Myth of Wild Africa: Conservation without Illusion* (Berkeley: University of California Press, 1996).

52. A recent tragic case in point is in Burma; see Adrian Levy, Cathy Scott-Clark, and David Harrison, "Save the Rhino, but Kill the People," *Manchester Guardian Weekly*, 30 March 1997, 5.

53. Bradley S. Klein, "Cultural Links: An International Political Economy of Golf Course Landscapes," in Randy Martin and Toby Miller, eds., *Sportcult* (Minneapolis: University of Minnesota Press, 1999), 211–26.

54. J. M. Jamil Brownson, *In Cold Margins: Sustainable Development in Northern Bioregions* (Missoula, Montana: Northern Rim Press, 1995).

55. Bruce Willems-Braun, "Buried Epistemologies: The Politics of Nature in (Post)Colonial British Columbia," *Annals of the Association of American Geographers* 87, no. 1 (1997): 3–31.

56. See, for example, Julian Simon, ed., *The State of Humanity* (Oxford: Blackwell, 1995); and Richard D. North, *Life on a Modern Planet: A Manifesto for Progress* (Manchester: Manchester University Press, 1995).

57. Botkin, *Discordant Harmonies.*

58. Thom Kuehls, *Beyond Sovereign Territory: The Space of Ecopolitics* (Minneapolis: University of Minnesota Press, 1996); R. B. J. Walker, "International Relations and the Concept of the Political," in Booth and Smith, *International Relations Theory Today*, 306–27.

8. ECOLOGY AND SECURITY STUDIES

1. F. O. Hampson and J. Reppy, eds., *Earthly Goods: Environmental Change and Social Justice* (Ithaca, N.Y.: Cornell University Press, 1996).

2. See Wolfgang Sachs, ed., *The Development Dictionary* (London: Zed, 1992); Wolfgang Sachs, ed., *Global Ecology* (London: Zed, 1993); and Sachs, *Planet Dialectics: Explorations in Environment and Development* (London: Zed, 1999).

3. Peter Taylor, *The Way the Modern World Works: World Hegemony to World Impasse* (London: Wiley, 1996), chapter 6; Richard Falk, *On Humane Governance: Toward a New Global Politics* (University Park: Pennsylvania University Press, 1995).

4. Timothy W. Luke, *Ecocritique: Contesting the Politics of Nature, Economy, and Culture* (Minneapolis: University of Minnesota Press, 1997); see also John S. Dryzek, *The Politics of the Earth: Environmental Discourses* (Oxford: Oxford University Press, 1997).

5. See Cara Stewart, "Old Wine in Recycled Bottles," paper presented at the annual meeting of the British International Studies Association, Leeds, December 1997; this is a criticism rejected by Matthew Paterson in the conclusion to his *Understanding Global Environmental Politics* (London: Macmillan, 2000).

6. Hence the juxtaposition of the two epigraphs to this chapter, one from an ecologist, the other from a social scientist; see Vaclav Smil, *Global Ecology: Environmental Change and Social Flexibility* (London: Routledge, 1993), and Barry Hindess, *Discourses of Power: From Hobbes to Foucault* (Oxford: Blackwell, 1996).

7. Gwyn Prins, "The Four-Stroke Cycle in Security Studies," *International Affairs* 74, no. 4 (1998): 781–808.

8. Bradley S. Klein, "Politics by Design: Remapping Security Landscapes," *European Journal of International Relations* 4, no. 3 (1998): 327–45. See also Mark Neocleous, "Against Security," *Radical Philosophy* 100 (2000): 7–14.

9. Ian Loader, "Thinking Normatively about Private Security," *Journal of Law and Society* 24, no. 3 (1997): 377–94.

10. Keith Krause, "Critical Theory and Security Studies: The Research Programme of 'Critical Security Studies,'" *Cooperation and Conflict* 33, no. 3 (1998): 298–333.

11. Steve Smith, "The Self-Images of a Discipline: A Genealogy of International Relations Theory," in Ken Booth and Steve Smith, eds., *International Relations Theory Today* (Cambridge: Polity, 1995), 1–37.

12. Michael Shapiro, *Violent Cartographies: Mapping Cultures of War* (Minneapolis: University of Minnesota Press, 1997).

13. I read Roxanne Lynn Doty's analysis in her *Imperial Encounters: The*

Politics of Representation in North-South Relations (Minneapolis: University of Minnesota Press, 1996) this way. Likewise, Michael Shapiro, *Violent Cartographies,* and Siba N'Zatioula Grovogui, *Sovereigns, Quasi Sovereigns, and Africans: Race and Self-Determination in International Law* (Minneapolis: University of Minnesota Press, 1996), offer important extensions of this point.

14. Craig Murphy and Thomas G. Weiss suggest that these themes are finally making their way into mainstream discussions of security studies in "International Peace and Security at a Multilateral Moment: What We Seem to Know, What We Don't, and Why," in Stuart Croft and Terry Terriff, eds., *Critical Reflections on Security and Change* (London: Frank Cass, 2000), 116–41.

15. Johan Galtung, "Violence, Peace, and Peace Research," *Journal of Peace Research* 6 (1969): 167–91; Johan Galtung, "A Structural Theory of Imperialism," *Journal of Peace Research* 8 (1971): 81–117.

16. A point made very clearly in a recent symposium on these matters; see Johan Eriksson, "Observers or Advocates? On the Political Role of Security Analysts," *Cooperation and Conflict* 34, no. 3 (1999): 311–30; and the discussion in Kjell Goldmann, "Issues, Not Labels, Please!" *Cooperation and Conflict* 34, no. 3 (1999): 331–33; Ole Waever, "Securitizing Sectors?" *Cooperation and Conflict* 34, no. 3 (1999): 334–40; Michael C. Williams, "The Practices of Security: Critical Contributions," *Cooperation and Conflict* 34, no. 3 (1999): 341–44; and Johan Eriksson, "Debating the Politics of Security Studies," *Cooperation and Conflict* 34, no. 3 (1999): 345–52.

17. Carlo Bonura, "The Occulted Geopolitics of Nation and Culture: Situating Political Culture within the Construction of Geopolitical Ontologies," in Gearóid Ó Tuathail and Simon Dalby, *Rethinking Geopolitics* (London: Routledge, 1998), 86–105.

18. Peter J. Taylor, *Modernities: A Geohistorical Interpretration* (Minneapolis: University of Minnesota Press, 1999).

19. John G. Ruggie, "Territoriality and Beyond: Problematizing Modernity in International Relations," *International Organisation* 47, no. 1 (1993): 139–74; and more generally, Ruggie, *Constructing the World Polity: Essays on International Institutionalization* (New York: Routledge, 1998).

20. Jon Barnett, *The Meaning of Environmental Security: Ecological Politics and Policy in the New Security Era* (London: Zed, 2001).

21. Nancy L. Peluso and Michael Watts, eds., *Violent Environments* (Ithaca, N.Y.: Cornell University Press, 2001).

22. Thom Kuehls, *Beyond Sovereign Territory: The Space of Ecopolitics* (Minneapolis: University of Minnesota Press, 1996).

23. Gustavo Esteva and Madhu Suri Prakesh, *Grassroots Post-Modernism: Remaking the Soil of Cultures* (London: Zed, 1998).

24. See, for instance, Graciela Chichilnisky and Geoffrey Heal, "Economic Returns from the Biosphere," *Nature* 391 (12 February 1998): 629–30.

25. See the classic statement on this theme in Vandana Shiva, "Conflicts of Global Ecology: Environmental Activism in a Period of Global Reach," *Alternatives* 19, no. 2 (1994): 195–207. More generally on the diversity of global perspectives, see Thomas W. Giambelluca and Ann Henderson-Sellers, eds., *Climate Change: Developing Southern Hemisphere Perspectives* (New York: Wiley, 1996); J. Gupta, *The Climate Change Convention and Developing Countries: From Conflict to Consensus?* (Dordrecht: Kluwer, 1997); and Tim O'Riordan and Jill Jager, eds., *Politics of Climate Change: A European Perspective* (London: Routledge, 1996).

26. A. Sen, *Poverty and Famines* (Oxford: Clarendon, 1981). See also Michael Watts, *Silent Violence: Food Famine and Peasantry in Northern Nigeria* (Berkeley: University of California Press, 1983).

27. A detailed discussion of these points is beyond the scope of this chapter, but see, for instance, Jonathan Crush, ed., *Power of Development* (London: Routledge, 1995); Majid Rahnema and Victoria Bawtree, eds., *The Post-Development Reader* (London: Zed, 1997); and James Scott, *Seeing like a State: How Certain Schemes to Improve the Human Condition Have Failed* (New Haven, Conn.: Yale University Press, 1998).

28. R. B. J. Walker, *Inside/Outside: International Relations as Political Theory* (Cambridge: Cambridge University Press, 1993).

29. D. B. Bobrow, "Complex Insecurity: Implications of a Sobering Metaphor," *International Studies Quarterly* 40 (1996): 435–50.

30. To add postcolonial and postdevelopment understandings would produce a security studies agenda very different from the regional variations in current debates; see P. B. Stares, ed., *The New Security: A Global Survey* (Tokyo: Japan Center for International Exchange, 1998).

31. John Mowitt, "In/Security and the Politics of Disciplinarity," in J. Weldes, M. Laffey, H. Gusterson, and R. Duvall, eds., *Cultures of Insecurity: States, Communities, and the Production of Danger* (Minneapolis: University of Minnesota Press, 1999), 359.

32. Patricia Molloy, "From the Strategic Self to the Ethical Relation: Pedagogies of War and Peace" (Ph.D. diss., University of Toronto, 1999).

33. François Debrix, *Re-envisioning Peacekeeping: The United Nations and the Mobilization of Ideology* (Minneapolis: University of Minnesota Press, 1999). Debrix cites Michel Foucault, *The Birth of the Clinic: An Archaeology of Medical Perception* (New York: Vintage, 1973).

34. Laurie Garrett, *The Coming Plague: Newly Emerging Diseases in a World out of Balance* (New York: Farrar Strauss, 1994).

35. Michael Hardt and Antonio Negri, *Empire* (Cambridge: Harvard University Press, 2000).

36. Mary Kaldor, *New and Old Wars: Organized Violence in a Global Era* (Stanford, Calif.: Stanford University Press, 1999).

37. Ulrich Beck, *Risk Society: Towards a New Modernity* (London: Sage, 1996); S. Lash, B. Szerszynski, and B. Wynne, eds., *Risk, Environment, and Modernity: Towards a New Ecology* (London: Sage, 1996).

38. Richard Wyn-Jones has also invoked Beck's "risk society" in his rather different formulations of a critical theory for security studies in *Security, Strategy, and Critical Theory* (Boulder, Colo.: Lynne Rienner, 1999).

39. Ulrich Beck, "World Risk Society as Cosmopolitan Society? Ecological Questions in a Framework of Manufactured Uncertainties," *Theory, Culture, and Society* 13 (1996): 1, 15.

40. See Gearóid Ó Tuathail, "Deterritorialised Threats and Global Dangers: Geopolitics and Risk Society," *Geopolitics* 3, no. 1 (1998): 17–31; and Ó Tuathail, "Understanding Critical Geopolitics: Geopolitics and Risk Society," in Colin S. Gray and Geoffrey Sloan, eds., *Geopolitics: Geography and Strategy* (London: Frank Cass, 1999), 107–24.

41. Barry Buzan, Ole Wæver, and Jaap de Wilde, *Security: A New Framework for Analysis* (Boulder, Colo.: Lynne Rienner, 1998).

42. M. D. Wolpin, "Third World Military Roles and Environmental Security," *New Political Science* 23 (1992): 91–120.

43. Commission on Global Governance, *Our Global Neighbourhood* (Oxford: Oxford University Press, 1995). For a critique of this viewpoint, see U. Baxi, "'Global Neighbourhood' and the 'Universal' Otherhood: Notes on the Report of the Commission on Global Governance," *Alternatives* 21 (1996): 525–49.

44. Paul Wapner, *Environmental Activism and World Civic Politics* (Albany: State University of New York Press, 1996).

45. John Vogler, *The Global Commons: Environmental and Technological Governance* (London: Wiley, 2000).

46. Karen Litfin, *Ozone Discourses: Science and Politics in Global Environmental Cooperation* (New York: Columbia University Press, 1994).

47. David Campbell, "Political Prosaics, Transversal Politics, and the Anarchical World," in Michael Shapiro and Hayward Alker, eds., *Challenging Boundaries: Global Flows, Territorial Identities* (Minneapolis: University of Minnesota Press, 1996).

48. John Vidal, "A Tribe's Suicide Pact," *Manchester Guardian Weekly,* 12 October 1997, 8–9. In general, see Al Gedicks, *Resource Rebels: Native Challenges to Mining and Oil Corporations* (Cambridge, Mass.: South End Press, 2001).

49. Patricia Mische, "Ecological Security and the Need to Reconceptualise Sovereignty," *Alternatives* 14, no. 4 (1989): 389–427.

50. Gearóid Ó Tuathail, "At the End of Geopolitics? Reflections on a

Plural Problematic at the Century's End," *Alternatives* 22 (1997): 35–56; Ó Tuathail, "Postmodern Geopolitics? The Modern Geopolitical Imagination and Beyond," in Ó Tuathail and Simon Dalby, *Rethinking Geopolitics* (London: Routledge, 1998), 16–38; John Agnew, *Geopolitics: Revisioning World Politics* (London: Routledge, 1998).

51. Gilles Deleuze and Felix Guattari, *A Thousand Plateaus: Capitalism and Schizophrenia* (Minneapolis: University of Minnesota Press, 1987); see also Thom Kuehls, *Beyond Sovereign Territory: The Space of Ecopolitics* (Minneapolis: University of Minnesota Press, 1996), chapter 2, on this point.

52. Michael Shapiro, "Moral Geographies and the Ethics of Post-Sovereignty," *Public Culture* 6 (1994): 479–502; R. B. J. Walker, "Social Movements/World Politics," *Millennium* 23, no. 3 (1994): 669–700; Roland Bleiker, *Popular Dissent, Human Agency, and Global Politics* (Cambridge: Cambridge University Press, 2000).

9. SECURING WHAT FUTURE?

1. Mustapha Pasha, "Security as Hegemony," *Alternatives* 21, no. 3 (1996): 287.

2. Betsy Hartmann, "Population, Environment, and Security: A New Trinity," in J. M. Silliman and Y. King, eds., *Dangerous Intersections: Feminist Perspectives on Population, Environment, and Development* (Boston: South End 1999), 1–23. See also Andrew Ross, "The Lonely Hour of Scarcity," *Capitalism, Nature, Socialism* 7, no. 3 (1996): 3–26.

3. Xenos, *Scarcity and Modernity* (London: Routledge, 1989).

4. See Steven Miles, *Consumerism as a Way of Life* (London: Sage, 1998).

5. See the classic discussion in Fred Hirsch, *Social Limits to Growth* (Cambridge: Harvard University Press, 1976); and in general, Roger Mason, *The Economics of Conspicuous Consumption: Theory and Thought Since 1700* (Cheltenham, U.K.: Elgar, 1998).

6. Nicholas Xenos, *Scarcity and Modernity*, 116.

7. Michel Chossudovsky, *The Globalization of Poverty: Impacts of IMF and World Bank Reforms* (London: Zed, 1997).

8. Michael Shapiro, "Images of Planetary Danger: Luciano Benetton's Ecumenical Fantasy," *Alternatives* 19, no. 4 (1994): 433–54; P. McHaffie, "Decoding the Globe: Globalism, Advertising, and Corporate Practice," *Environment and Planning D: Society and Space* 15, no. 1 (1997): 73–86.

9. Steven Yearley, *Sociology, Environmentalism, Globalization: Reinventing the Globe* (London: Sage, 1996).

10. This is the case in Britain; see Alan Hallsworth, Rodney Tolley, Colin

Black, and Ian Cross, "Fear and the Car as Haven: Parental Justifications for Car-Use on Trips to School," working paper, Centre for Alternative and Sustainable Transport, Staffordshire University, 1997.

11. See Heather Smith, "Human (In)security and Canadian Climate Change Policy," *Environment and Security* 4 (2000): 89–101.

12. According to the United Nations statistics used in the compilation of the human security agenda, slightly over 12 American males are murdered per 100,000 of population per annum. The comparative figures per year per 100,000 are that about 120 women report rapes and nearly 1,400 injuries occur from road accidents (United Nations Development Program, *Human Development Report 1994* [New York: Oxford University Press, 1994], 30).

13. Nonetheless, and contrary to many on the political left, at least Michael Hardt and Antonio Negri see potentials in the resistance to what they see as the emergence of empire in contemporary times. See Hardt and Negri, *Empire* (Cambridge: Harvard University Press, 2000).

14. Escobar, "Constructing Nature," in Richard Peet and Michael Watts, eds., *Liberation Ecologies: Environment, Development, Social Movements* (New York: Routledge, 1996), 53.

15. Robert Zubrin, *The Case for Mars: The Plan to Settle the Red Planet and Why We Must* (New York: Free Press, 1996). The extended argument for going beyond Mars and the asteroids is laid out in Robert Zubrin, *Entering Space: Creating a Spacefaring Civilization* (New York: Tarcher/Putnam, 1999). See also "Special Report: Sending Astronauts to Mars," *Scientific American* 282, no. 3 (2000): 40–63.

16. See R. P. Turco, O. B. Toon, T. P. Ackerman, J. B. Pollack, and C. Sagan, "Nuclear Winter: Global Consequences of Multiple Nuclear Explosions," *Science* 222 (1983): 1283–92.

17. Zubrin, *The Case for Mars,* 248.

18. Kim Stanley Robinson, *Red Mars* (New York: Bantam, 1993); *Green Mars* (New York: Bantam, 1994); *Blue Mars* (New York: Bantam, 1996).

19. Barry Buzan, Ole Wæver, and Jaap de Wilde, *Security: A New Framework for Analysis* (Boulder, Colo.: Lynne Rienner, 1998).

20. James Scott, *Seeing like a State: How Certain Schemes to Improve the Human Condition Have Failed* (New Haven, Conn.: Yale University Press, 1998).

21. On the irony of central planning to solve global capitalist crisis, see Mohameden Ould-Mey, "The New Global Command Economy," *Environment and Planning D: Society and Space* 17, no. 2 (1999): 155–80.

22. This paragraph and the one following are a distillation from the crucial debate in A. Dwight Baldwin Jr., Judith de Luce, and Carl Pletsch, eds., *Beyond Preservation: Restoring and Inventing Landscapes* (Minneapolis: University of Minnesota Press, 1994).

23. This was one of Jeremy Rifkin's concerns in his *Biospheric Politics: A New Consciousness for a New Century* (New York: Crown, 1991).

24. Bruce Braun and Noel Castree, eds., *Remaking Reality: Nature at the Millennium* (London: Routledge, 1998).

25. John Agnew, *Geopolitics: Revisioning World Politics* (London: Routledge, 1998).

26. Zubrin, *The Case for Mars*, 248–49.

27. The cancer metaphor is used in Van B. Weigel, *Earth Cancer* (Westport, Conn.: Praeger, 1995), and turned into a mode of diagnosing the ills of capitalism in John McMurtry, *Cancer Stage of Capitalism* (London: Pluto, 1999).

28. The Centre for Science and Environment, "Statement on Global Environmental Democracy," *Alternatives: Social Transformation and Humane Governance* 17, no. 2 (1992): 261–79; Pratap Chatterjee and Matthias Finger, *The Earth Brokers: Power, Politics, and World Development* (London: Routledge, 1994).

29. Vaclav Smil, *Global Ecology: Environmental Change and Social Flexibility* (London: Routledge, 1993),

30. For a recent popularization linked to attempts to rethink business practices, see Paul Hawken, Amory Lovins, and L. Hunter Lovins, *Natural Capitalism: Creating the Next Industrial Revolution* (New York: Little Brown, 1999). This is, of course, not a very new argument, neither is the case that reducing dependency on oil imports makes sense as a national security stategy. See Amory Lovins and L. Hunter Lovins, *Brittle Power: Energy Strategy for National Security* (Andover, Mass: Brick House, 1982).

31. R. B. J. Walker, "The Subject of Security," in Keith Krause and Michael C. Williams, eds., *Critical Security Studies: Concepts and Cases* (Minneapolis: University of Minnesota Press, 1997), 78.

32. R. T. Watson, M. Zinyowera, and R. H. Moss, eds., *Climate Change 1995: Impacts, Adaptations, and Mitigation of Climate Change: Scientific-Technical Analyses* (Cambridge: Cambridge University Press, 1996).

33. Manuel Castells, *The Rise of the Network Society*, vol. 1 of *The Information Age* (Oxford: Blackwell, 1997).

34. John Agnew, "Mapping Political Power beyond State Boundaries: Territory, Identity, and Movement in World Politics," *Millennium: Journal of International Studies* 28, no. 3 (1999): 499–521.

35. Richard Falk formulates these terms as "Predatory Globalization" in *Predatory Globalization: A Critique* (Cambridge, England: Polity Press, 1999).

36. Etel Solingen, *Regional Orders at Century's Dawn: Global and Domestic Influences on Grand Strategy* (Princeton, N.J.: Princeton University Press, 1998).

37. Michael Ignatief, *Virtual War: Kosovo and Beyond* (London: Viking, 2000).

38. Mary Kaldor, *New and Old Wars: Organized Violence in a Global Era* (Stanford, Calif.: Stanford University Press, 1999).

39. David Campbell and Michael J. Shapiro, eds., *Moral Spaces: Rethinking Ethics and World Politics* (Minneapolis: University of Minnesota Press, 1999). On animals and ethics, see Jennifer Wolch and Jody Emel, eds., *Animal Geographies: Place, Politics, and Identity in the Nature-Culture Borderlands* (London: Verso, 1998).

40. Allen Hammond, *Which World: Scenarios for the 21st Century* (Washington, D.C.: Island Press, 1998); Mike Davis, *Ecology of Fear: Los Angeles and the Imagination of Disaster* (New York: Vintage, 1999).

41. Peter J. Taylor, *Modernities: A Geohistorical Interpretration* (Minneapolis: University of Minnesota Press, 1999).

42. Winona LaDuke attributes this teaching to the Six Nations Iroquois Confederacy, in *All Our Relations: Native Struggles for Land and Life* (Cambridge, Mass.: South End, 1999), 198.

43. See Marcel Wissenburg, *Green Liberalism: The Free and the Green Society* (London: University College London Press, 1998), on just how difficult these dilemmas are.

44. Security scholars might also note that Zubrin argues in *The Case for Mars,* in parallel with recent Hollywood movies, that space travel capabilities, rather than weapons, offer much better prospects for fending off any asteroids that might in future threaten to collide catastrophically with the earth.

45. I am obviously ignoring the possibility that humans are merely the means to the emergence of a more powerful machinic "life" form; see Bill Joy, "Why the Future Doesn't Need Us," *Wired* (April 2000): 238–62.

Index

SIMON DALBY is professor of geography and political economy at Carleton University in Ottawa. He is coeditor of *The Geopolitics Reader* and *Rethinking Geopolitics*.